Mobile Computing

Herausgegeben von
Ch. Kittl, Graz, Österreich

Das Mediennutzungsverhalten der Menschen hat sich in den letzten Jahren stark intensiviert und gewandelt. Die Kombination aus der raschen Verbreitung mobiler Endgeräte und der Einführung neuer Dienste jenseits der reinen Sprachtelefonie führt zu einer Fülle neuer Möglichkeiten sowohl für Konsumenten als auch für Anbieter von Mobilfunkdiensten. Im Rahmen der Schriftenreihe werden herausragende Arbeiten aus dem breiten Forschungsfeld des Mobile Computing publiziert. Darin werden unter Einhaltung wissenschaftlicher Methoden Problemfelder aus der Praxis bearbeitet und Lösungsansätze vorgestellt. Das Hauptaugenmerk liegt neben der Entwicklung technologischer Artefakte auf der Ausgestaltung ökonomisch sinnvoller Geschäftsmodelle und der Nutzerakzeptanz solcher Systeme.

Herausgegeben von
DI Dr. Christian Kittl
evolaris next level GmbH,
Graz, Österreich

Gerhard Schall

Mobile Augmented Reality for Human Scale Interaction with Geospatial Models

The Benefit for Industrial Applications

Foreword by Professor Dr. Dieter Schmalstieg

 Springer Gabler

RESEARCH

Gerhard Schall
Graz, Austria

Dissertation Graz University of Technology, Austria, 2011

ISBN 978-3-658-00196-4   ISBN 978-3-658-00197-1 (eBook)
DOI 10.1007/978-3-658-00197-1

The Deutsche Nationalbibliothek lists this publication in the Deutsche Nationalbibliografie; detailed bibliographic data are available in the Internet at http://dnb.d-nb.de.

Library of Congress Control Number: 2012949282

Springer Gabler
© Springer Fachmedien Wiesbaden 2013

Printed on acid-free paper

Springer Gabler is a brand of Springer DE. Springer DE is part of Springer Science+Business Media.
www.springer-gabler.de

Foreword

Augmented Reality (AR) has recently become recognized as an emerging new medium. A broad audience today is interested in AR applications such as personal navigation, marketing or games. However, the history of AR applications in an industrial context goes further back. This field of application is currently reaching a turning point that marks the productive deployment of AR in the field.

You are holding the dissertation of Dr. Gerhard Schall in your hands, which deals with the use of AR systems for industrial applications in wide outdoor environments. Such an AR system requires at least three significant components: a model of the environment in which the system should be deployed, a real-time capable method for tracking the human user and an ergonomically acceptable mobile hardware setup. These three are the core topics of this book.

Maybe the most important of the three topics is the geospatial model that is suitable for AR applications. This implies that the model allows both localization from and visualization of the model data, which requires novel structures for the model data. Moreover, the model must be measured from scratch or converted from existing sources; both cases require appropriate processing method. Ultimately, the AR system must support the user in a better way than hitherto possible with conventional maps. A key achievement of the work described in this book is the systematic development of technology that addresses these requirements.

The first part of the book is concerned with two kinds of model creation. Models of large indoor environments are acquired using a mobile robot equipped with a computer vision system. Data from outdoor underground infrastructure databases is transcoded into 3D models.

The second part of the book describes the design and crafting of handheld AR systems relying on tablet computers. The AR device is held with both hands and has suitable ergonomic properties for outdoor work of engineers.

The third part of the book deals with outdoor 3D tracking. A hybrid tracking system is presented, which integrates GPS, compass and inertial sensor with a visual orientation tracker. Using sensor fusion, improved robustness and precision can be achieved.

Overall, this book points into new directions concerning outdoor AR for industrial applications. It can be assumed that AR is soon going to be an important medium for presenting geospatial information in-situ, i.e., directly at the task location where engineers work.

Prof. Dr. Dieter Schmalstieg
Head of the Institute for
Computer Graphics and Vision
Graz University of Technology

Acknowledgements

I gratefully thank my family for all the continuous support over all the years. Furthermore, my thank goes to current and former colleagues from the Institute for Computer Graphics and Vision at the University of Technology Graz and our industrial partners for their support, collaboration, help, comments and critique. I especially express my gratitude towards Dieter Schmalstieg, who introduced me into this research field and provided the necessary freedom to do my research, and Gerhard Reitmayr, who contributed in manifold ways with his ideas and insights. Joseph Newman and Erick Mendez deserve special mentioning for their long-time collaboration and doing many challenging live demos together. I also thank my second supervisor Tobias Höllerer from the University of California, Santa Barbara, for the support and guidance and last but not least Franz Leberl for his outstanding visions and leadership.

Why do I write about augmented reality topics? Probably the major reason is my intense interest in the potential of AR and the joy of exploring new technologies. Even more important, I hope that mobile and handheld augmented reality can be further enhanced and find adoption in real-world scenarios. I foresee new ways of interacting using augmented reality as a user interface as well as new and changing fields of application.

The topic of this thesis is tied to several fields of research thus it is highly interdisciplinary. Topics from wearable computing, pervasive and ubiquitous computing will be mixed with more specialized fields like virtual and augmented reality technology. There are simply too many topics related to the theme of this thesis – this makes it impossible to explore all possible topics fully. Hence, wherever possible, sources for further reading are provided to overcome possible gaps and to serve the reader when there is more interest in a specific topic.

Dr. Gerhard Schall
Institute for Computer Graphics and Vision
Graz University of Technology

Abstract

Augmented reality (AR) is intended to present new and more meaningful interfaces to users. At the same time there is a trend towards more realistic representation of real-world information as computing devices are becoming more pervasive than ever. Mobile AR has left the research lab and demonstrated its educational, social and economical potential. This thesis contributes to the research with several novel tracking approaches and a set of tools for creating content for AR applications.

One necessity for successful AR application is accurate and robust tracking of the user's position and orientation (pose). This thesis covers several approaches to achieving robust and accurate tracking for AR. A marker-based hybrid tracking system using ultra-wide-band and inertial sensors for indoor environments is described. Furthermore, the thesis presents a multi-sensor fusion approach for combining differential data from the global positioning system, inertial sensor data as well as pose estimations with a visual tracker.

Another integral part of an AR application is the computer graphic content. Only non-manual modeling approaches can fulfill the need for larger areas and more complex content. Data sources such as geospatial information systems can be exploited for the creation of content. Following this vision, this thesis presents a content modeling approach describing a transcoding pipeline which generates models for AR by taking advantage of the rich data stored in geospatial databases. The models contain visual and non-visual information for the dual purpose of visualization and tracking.

There is a close relation between the content and the applied tracking approach. In particular, the coordinate systems of both methods need to fit together to achieve properly registered overlays. The approaches developed are applied in several mobile AR applications, among them an industrial application for registered visualization of subsurface infrastructure. Various tools, for example a virtual redlining annotation feature, are described and expert interviews are provided. In addition, evaluations from real-world test sites are presented. The thesis concludes with a summary of and reflection on the status quo, including a road map of open issues for further research.

Kurzfassung

Augmented Reality (AR) ist ein User-Interface-Paradigma, das virtuelle und reale Informationen verschmelzen läßt. Die Überlagerung der realen Umgebung mit virtuellen Informationen mit interaktiven Frameraten wirft eine Reihe von Themen auf. Diese beinhalten unter anderem die Forschungsgebiete der Erzeugung von 3-D-Modellen für AR und der Positions- und Orientierungsbestimmung (auch Tracking genannt) des mobilen Benutzers.

Um diese Probleme zu behandeln, befasst sich diese Arbeit mit der hochgenauen, robusten und stabilen Positions- und Orientierungsbestimmung. Für den Einsatz im Inneren von Gebäuden sowie im Freien wird eine Reihe von hybriden Techniken vorgestellt, die unterschiedliche Sensoren, wie Intertialsensoren, Magnetometer, Ultrawideband, differentielle GPS-Empfänger und Kameras intelligent integrieren. Alle präsentierten Techniken sind geeignet, ihren jeweiligen Zweck erfolgreich zu erfüllen.

Weiters befasst sich diese Arbeit mit der Erzeugung von größeren und komplexen 3-D-Modellen für AR, die auch kontextuelle Informationen enthalten. In diesem Zusammenhang ist eine manuelle Erzeugung der Modelle nicht zielführend. Daher präsentiert diese Arbeit einen Modellierungsansatz basierend auf einem Transcoding-Verfahren, welches auf effiziente und automatische Weise prozedurale 3-D-Modelle aus Daten von geographischen Informationssystemen generieren kann. Die erzeugten Modelle enthalten neben der Geometrie auch kontextuelle Informationen, die für interaktive Visualisierungen als auch für die Positions- und Orientierungsbestimmung verwendet werden.

Zusätzlich müssen die Koordinatensysteme der 3-D-Modelle und Trackingsysteme zueinander registriert werden, um die virtuellen Modelle passend der realen Welt in 3-D überlagern zu können. Die vorgestellten Methoden wurden in mehreren AR-Hardware-Prototypen, welche eigens dafür experimentell entwickelt wurden, eingesetzt. Am Beispiel der AR-Visualisierung von unterirdischer Leitungsinfrastruktur im industriellen Umfeld werden mehrere Werkzeuge zur Visualisierung und mobilen Benutzer-Interaktion vorgestellt. Das Potential von mobiler AR wird in Interviews mit Experten und Außendienstmitarbeitern, die im Rahmen von Feldtests und Benutzerstudien die AR-Prototypen getestet haben, bestätigt.

Contents

Foreword v

Acknowledgements vii

Abstract ix

Kurzfassung xi

Contents xiii

List of Figures xvii

List of Tables.. xxiii

1. Introduction... 1
    1.1 Augmented reality 2
    1.2 Ubiquitous computing 4
    1.3 Problem statement.. 6
    1.4 Hypotheses.. 7
    1.5 Contribution 8
    1.6 Collaboration statement.. 10
    1.7 Organization... 16

2. Background and related work 19
    2.1 Augmented reality displays.. 19
    2.2 Mobile augmented reality.. 20
    2.3 Geospatial models 24
    2.4 Pose tracking 28
    2.5 Discussion... 31

3. Requirements 33
    3.1 Sensor reference frame 33
    3.2 Data reference frame 34
    3.3 Global reference frame... 35
    3.4 Augmented reality models.. 36
    3.5 System design considerations.. 37

4. Interactive geospatial models for augmented reality 39
    4.1 Manual surveying 39
    4.2 Semi-automatic surveying... 44

4.3   Example application 49

4.4   Transcoding pipeline.. 53

        4.4.1 Transcoding process 53

        4.4.2 Anatomy of the geospatial infrastructure 57

        4.4.3 Transcoding trade-off analysis 62

        4.4.4 Limitations... 65

4.5   Discussion.. 65

5.   Hardware setups for augmented reality................................... 67

5.1   Requirements.. 68

5.2   UMPC-based setups... 69

5.3   Vesp´R setup 71

5.4   Evaluation of Vesp´R setup 72

        5.4.1 Mobile computing developers 73

        5.4.2 Mixed user group... 75

        5.4.3 Field worker interview 77

        5.4.4 Management level feedback 79

        5.4.5 Evaluation summary 79

5.5   POMAR-3D setup 80

5.6   Tablet PC-based setup 83

5.7   Discussion.. 84

6.   Pose tracking 85

6.1   Global pose estimation using multi-sensor fusion 85

6.2   Position tracking... 87

6.3   Orientation tracking 89

        6.3.1 Gyroscope measurement model 90

        6.3.2 Magnetometer measurement model 91

        6.3.3 Accelerometer measurement model.......................... 91

6.4   Visual tracking approach.. 92

6.5   Fusion of attitude with visual tracking 94

6.6   Position using differential GPS... 97

6.7   Position using real-time kinematic GPS................................. 100

6.8   Attitude Kalman filter 106

6.9   Visual tracking 110

6.10 Combination of attitude with visual tracking 112

6.11 North-centered orientation tracking..114

6.11.1 Orientation estimation and sensor fusion 115

6.11.2 Kalman filter setup 117

6.11.3 Results...118

6.12 Discussion..127

7. Applications – AR visualization and interaction in civil engineering 129

7.1 Concept 130

7.1.1 Inspection toolset..133

7.1.2 Excavation tool 134

7.1.3 Metadata Querying tool..135

7.1.4 Filtering tool...136

7.2 Snapshot tool 136

7.3 Interactive redlining toolset 136

7.3.1 Annotating the geospatial model..137

7.3.2 Surveying in the geospatial model 138

7.3.3 Interactive validation of object placement..................................138

7.4 Verification toolset...139

7.4.1 Visualization of abstract information 139

7.4.2 Verification of abstract information...140

7.5 Evaluation results 140

7.5.1 Evaluation procedure...142

7.5.2 Digital terrain model...143

7.5.3 Virtual redlining..143

7.5.4 Verification of abstract information...146

7.6 Role of AR in field information systems 146

7.7 Discussion..149

8. Conclusions...151

8.1 Reflection 151

8.2 Road map 154

Appendix 155

Bibliography 163

List of Figures

Figure 1: The reality-virtuality continuum of Milgram. 3

Figure 2: Milgram-Weiser diagram of (Newman et al., 2007). 6

Figure 3: The first head-mounted display by Ivan Sutherland. 19

Figure 4: Evolution of mobile AR systems. Hardware setups range from backpack systems to handheld computers. 21

Figure 5: Mobile AR systems. 21

Figure 6: Going out. (Left) A user operating a handheld augmented reality unit tracked in an urban environment. (Middle) Live shot showing the unit tracking a building. (Right) Screenshot from a pose close to the left images with overlaid building outline. 22

Figure 7: A user operating a smart phone using an AR application for visualizing labels registered on the environment. 23

Figure 8: Manually generated 3D model of a building floor. The model includes a corridor and few dozens rooms. 25

Figure 9: 3D city model and building models in various detail based on CityGML. 27

Figure 10: Table top visualization. 29

Figure 11: Coordinate systems of reference frames. 34

Figure 12: Reference frames. 35

Figure 13: Manual surveying and modeling workflow. 41

Figure 14: Spatial relationships involved in the bridging of adjacent surveying areas (top). Graph of interrelationships (bottom). 43

Figure 15: Indoor building model. 44

Figure 16: Semi-automatic surveying. 48

Figure 17: Superimposed map of the markers reconstructed from the robot and the map of markers measured with the total station 48

Figure 18: Two-dimensional floor plan measured by the PeopleBot robot. 48

Figure 19: Indoor tracking setup. 51

Figure 20: Hybrid tracking approach for indoor environments. 51

Figure 21: Handheld AR user performing navigation task. 52

Figure 22: Navigating towards destination "corridor". 52

Figure 23: 2D map vs. 3D model of urban area... 53

Figure 24: Transcoding pipeline. 54

Figure 25: Transcoding output formats. 59

Figure 26: Curve geometry... 59

Figure 27: Trench with different opening angles. 61

Figure 28: 3D models representing GIS symbols.. 61

Figure 29: 3D model of an urban area in Salzburg 63

Figure 30: 3D model of an urban area in Vienna. 64

Figure 31: 3D model of an urban area in Graz in Sandgasse/Inffeldgasse. 64

Figure 32: Unconventional indoor AR setup. 69

Figure 33: Indoor AR setup... 70

Figure 34: Outdoor AR setup named Vesp´R. 71

Figure 35. A participant testing the Vidente application exhibited during the Ubicomp 2007 conference. 74

Figure 36: Evaluation results.. 76

Figure 37: Expert field worker.. 77

Figure 38: Pomar-3D, outdoor AR setup.. 81

Figure 39: Pomar-3D, user holding outdoor AR setup. 81

Figure 40: Differential GPS tracking module... 82

Figure 41: Tablet PC-based AR setup.. 83

Figure 42: Multi-sensor fusion system architecture. 86

Figure 43: State machine SM {Σ, S_0, S1, S2, S3, S4, δ}. 95

Figure 44: 3D model of the test site. 96

Figure 45: GPS measurements.. 98

Figure 46: Position estimates along a path using the Position Kalman filter. 98

Figure 47: AR overlay. 99

Figure 48: Rooftop test scenario. 100

Figure 49: GPS accuracy measurements of (a) northing, (b) easting and (c) height using the L1/L2 differential GPS receiver. 101

Figure 50: Error distribution of the overall re-projection error. 102

Figure 51: AR view of a re-projected physical reference point on the pavement......... 103

Figure 52: Histogram of re-projection errors for easting direction. 104

Figure 53: Histogram of re-projection errors for northing direction. 104

Figure 54: AR view with superimposed enclosures and base point of the building corner and a capping registered in 3D.. 105

Figure 55: Attitude during situation (a): Rotation of user, no drift (up: inertial sensor, down: Kalman filter attitude)... 107

Figure 56: Attitude during situation (b): Rotation of user, transient drift (up: inertial sensor, down: Kalman filter attitude)... 108

Figure 57: Attitude during situation (c): Rotation of user, permanent drift (left: inertial sensor, right: Kalman filter attitude). 109

Figure 58: Comparison of orientation estimates during start up phase and rotation of inertial sensor (yaw). 111

Figure 59: Screenshot showing the performance of inertial and visual tracking by visualizing building wireframe models. 111

Figure 60: Map of the outdoor environment created by the visual panorama tracker. 112

Figure 61: Comparison of yaw of the Kalman filtered inertial sensor with yaw of the visual panorama tracker under various conditions. 113

Figure 62: Reference systems. 116

Figure 63: Test area. 119

Figure 64: Test setup. 120

Figure 65: Errors (in degrees) to the north for the sequences recorded on the tablet PC (top) and on the phone (bottom). 122

Figure 66: Error distribution from the tablet PC (top) and for the phone (bottom), for both the sensors and the hybrid tracker. 123

Figure 67: Test sequence for the tablet PC. 124

Figure 68: Test sequence for the phone. 124

Figure 69: Plot of heading, pitch and roll for a free-hand movement of the tablet PC between two reference points. 126

Figure 70: AR visualization of GIS data. (Left) User with mobile AR system. (Right) Users view of very simple visualization of geospatial data. 131

Figure 71: User with mobile outdoor prototypes. 131

Figure 72: Screenshot of a working outdoor prototype of the project Vidente. 132

Figure 73: Smart Vidente system architecture. 132

Figure 74: Image-based ghostings. 134

Figure 75: AR view showing underground pipes well registered in 3D. 135

Figure 76: Excavation tool and Metadata querying tool. 135

Figure 77: Snapshot of the augmented live video. 136

Figure 78: Conventional redlining feature in 2D (left). Visualized in a conventional geographic information system (GE Smallworld™). (Right) AR redlining feature shown in 3D. 137

Figure 79: Noise protection barrier to be erected alongside a railroad track. 139

Figure 80: Conventional exocentric view at land register and underground services data as available in two-dimensional GIS visualizations (Graz Geodatenserver) (left). (Right) Egocentric view at land register. 140

Figure 81: The workflow of an inspection task on the left follows the method used to identify an underground object (left) using a 2D map and (right) utilizing a 3-dimensional AR visualization. 147

Figure 82: 2D map showing GIS features. 148

Figure 83: 3D AR visualization showing surveyed above ground features and subsurface features. 148

Figure 84: AR view with superimposed enclosures and base point of the building corner registered in 3D. The surface reference features are covered by winter snow. 150

Figure 85: Centimetre grid used for accuracy experiments. 159

Figure 86: Feature node in GeographyML format. 160

Figure 87: Open Inventor node of a pipe feature containing the subnodes for properties and geometry. 160

Figure 88: Open Inventor node of a pipe feature containing the subnodes for properties and geometry. 161

List of Tables

Table 1: Transcoding results of three different urban areas. 65

Table 2: State transition table. Shows which conditions induce a transition to which state. 95

Table 3: GPS accuracy measurements using GPS and DGPS. 97

Table 4: GPS accuracy measurements of the L1/L2 differential GPS receiver. 97

Table 5: Error in the overall re-projection. 102

Table 6: Attitude during test of relative angular accuracy. 106

Table 7: Yaw measurements of visual tracker. 113

Table 8: Angle (in degrees) to magnetic and true north of reference points 1-8, as seen from the reference point RP. 119

Table 9: Mean and standard deviation (in degrees) for the error of the sensors, the vision tracker and the hybrid tracker from both the tablet PC and the phone measurements. 125

Any sufficiently advanced technology
is indistinguishable from magic.

Arthur Clarke
Science fiction author

1. Introduction

People are ever more connected and can travel between locations while having real-time access to information sources, much of it having a spatial component. Anytime and anywhere, access to data is in demand by the wealth of nomadic users. Liberating users from indoor-based PCs and physical network connections opens new opportunities to allow geospatial information access in many real-world situations, thus revolutionizing how users interact with the world and surrounding environment through the use of handheld devices.

The goal of "anytime, anywhere" is to allow a roaming user to access information on demand. New methods which specifically accommodate user mobility will be required. Methods that were suitable for visual representation on PCs are not suitable for most mobile situations anymore where display screens are much smaller and bandwidth capacities are limited. Consequently, there is the need for new context-sensitive representations of geospatial information and consistent database access.

In field settings available technologies allow only for limited visual representations of geospatial information. Anyhow, a visual display is still the most appropriate display type for accessing geospatial data. More natural and intuitive user interfaces are needed to meet the new demands. Integrating mobile displays supporting natural mechanisms for interacting with map-like representations and augmented reality technologies poses a variety of technology and HCI challenges. To allow the user to acquire and use geospatial information in the field will also require focusing on spatial interaction issues. Here, efficient visualization techniques of geospatial models are vital. To meet this challenge there is existing work on map generalization that is an important conceptual base. Anyhow, augmented reality demands real-time generation of dynamically changing representations, which is still not solved today.

Classically, mobile augmented reality requires both detailed geospatial models describing the fixed world and accurate tracking support to register the user's location with that data. As geospatial data gets widespread, mobile augmented reality applications will become more important. Consider for example what benefits it might have for society if firefighters could look at a burning house and see important context-relevant information superimposed over on site. This could greatly help in fulfilling the firefighter's tasks more quickly.

Mobile augmented reality deals with more than only visualization and spatial interaction, going well beyond those already discussed. How can the system determine the data which should be superimposed? For example, point-and-click offers one suitable approach which allows spatial interaction with objects using a visual pointer or inertial sensor information, mixing the real and virtual world. The user should be enabled to point at an object to get access to identify the object and furthermore receive useful information about it.

Following this roadmap, various research problems must be solved before this vision becomes reality. The user's orientation and location (pose) must be determined very precisely. Among others, this will require real-time tracking, and a virtual 3D model of the environment. Moreover, the interface must support the user in achieving his task. The user interface must be as simple as possible, but not simpler. A user could request information about a specific geospatial object by pointing at it and gain information about it.

The vision of users roaming all over large spaces could become reality by future interdisciplinary research. The user could then be supported with additional information delivered by pervasive computer infrastructures depending on the current context and situation.

1.1   Augmented reality

In 1968 Sutherland proposed a new "Head-Mounted Three-Dimensional Display" superimposing computer generated models on the real environment (Sutherland, 1968). He believed that the ultimate display might give a better understanding of our own natural world. In 1997 Azuma defined augmented reality as the extension of a user's perception with virtual information. It has three main characteristics: combining real and virtual elements, being interactive in real-time and being registered in 3D (Azuma, 1997). This definition incorporates non-visual augmentation (e.g. audio AR) as well as mediated reality environments where a part of reality is replaced rather than augmented with computer-generated information. Exploiting these features, AR offers various new approaches and interfaces, especially for geospatial information. The spatial information can be directly displayed "on the spot", and the interaction can take place in a simple and intuitive way (Azuma 1997). Figure 1 shows another definition of augmented reality by using a reality-virtuality continuum (Milgram & Kishino, 1994). In contrast to virtual reality (VR), which completely immerses a user in a computer-generated environment,

AR aims at adding information to the users view and thereby allows experiencing both real and virtual information at the same time.

AR has close connections to the fields of VR and *mixed reality* (MR), where the virtual augments the real, and *augmented virtuality* (AV). However to better understand the relationships between these fields, the reality-virtuality continuum is helpful. Summarizing, the MR continuum describes the concept that there is a continuous scale between the completely virtual and the completely unmodified reality. The continuum therefore encompasses all possible variations and compositions of real and virtual objects. This thesis relies on the definitions of augmented reality of Milgram, Kishino and Azuma.

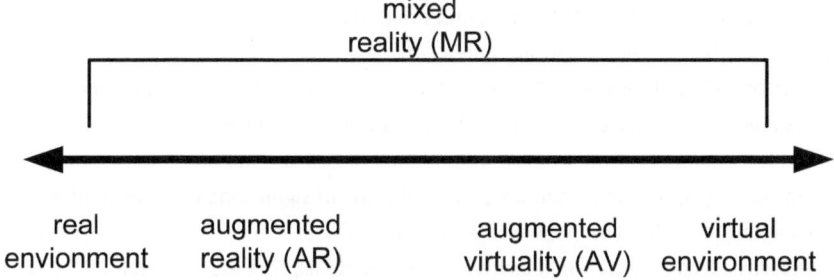

Figure 1: The reality-virtuality continuum of Milgram. It is a continuous scale ranging between the completely real (left), a mixed reality (middle) and the completely virtual (right) with a breakdown of the mixed reality segment.

AR systems are being introduced in industrial, commercial, medical and scientific markets for a variety of tasks such as computer-aided surgery and assisting in complex repair tasks (in airplanes, for instance). AR is attractive here for several reasons:

- It permits work to be done without having to look back and forth between the subject and reference material (such as a manual or medical imaging results).

- It can make complex three-dimensional tasks more easily understandable (and less prone to error) by providing more information overlaid onto the subject.

- AR can help with visualization and navigation of the highly complex data used in fields such as biotechnology research and development.

1.2   Ubiquitous computing

In 1991 Mark Weiser coined the term **ubiquitous computing** (ubicomp) and wrote that the most profound technologies are those that disappear. They weave themselves into the fabric of everyday life until they are indistinguishable from it. The Oxford English Dictionary ("Oxford Dictionaries Online - English Dictionary and Language Reference" 2010) gives the following explanations for the terms ubiquitous and pervasive.

> **Ubiquitous:** present, appearing, or found everywhere
> **Pervasive:** spreading widely throughout an area

Weiser stated that most of the computers that participate in embodied virtuality will be invisible in fact as well as in metaphor. These machines and others will be interconnected in a ubiquitous network. Weiser found two issues of crucial importance: *location* and *scale*. Little is more basic to human perception than physical juxtaposition, and so ubiquitous computers must know where they are. If a computer merely knows what room (or space) it is in, it can adapt its behavior in significant ways without requiring even a hint of artificial intelligence (Weiser, 1991).

Recent advances in sensors, embedded microsystems, wireless communications and the like have led to the evolution of the next generation of distributed computing platforms was described by Weiser two decades ago. Ubiquitous computing (also referred to as pervasive computing) moves processing and communication technology beyond the personal computer to everyday devices such as key chains, cars, homes and the human body. These everyday devices interconnect as a ubiquitous network of intelligent devices that cooperatively and autonomously collect, process and transport information in order to adapt to the context and activity. Ubiquitous computing seeks to provide supportive, proactive and self-tuning environments and devices to seamlessly augment a person's knowledge and decision making ability. Direct user interaction should be minimal. For example, the user could access and interact with information and services anytime and from anywhere in the world. The user doesn't need to carry around devices containing his or her information as ubiquitous environments will automatically serve as information storage that can be accessed by roaming users. Today *cloud computing* provides large central information storage and supports the realization of Weiser´s vision.

Moreover, ubicomp enables devices to be aware of their own surroundings (e.g. their position, orientation and environment) and thus sensing the presence of the user. When devices become more aware, they can be more responsive and seem smarter. Although the ubiquitous computing revolution has already begun to affect our lives in many ways, there are numerous challenges ahead as mobile and handheld devices become widespread. The trend towards ubiquitous computing is driving research into ever more natural forms of HCI (Abawajy, 2009).

The fields of wearable computing, augmented reality and ubiquitous computing are in principle highly convergent as they all promise a utopian future in which there is a level of integration such that users can intuitively perceive and interact with their environment. For example, sentient AR (Newman, Ingram, & Hopper, 2001) envisions that users can roam freely within buildings and access location-depended services. In this context, at AT&T Laboratories Cambridge researchers provide users with AR services using data from an ultrasonic tracking system called the Bat system which has been deployed building-wide.

Weiser stated that ubiquitous computing is roughly the opposite of VR. However when one considers that VR is merely at one extreme of the reality-virtuality continuum postulated by Milgram, then one can see that ubicomp and VR are not strictly opposite one another but rather orthogonal as described and illustrated herein. This new dimension was named the *Weiser's continuum* by Newman and would have ubicomp at one extreme and the concept of terminal-based computing at the other. The terminal is the antithesis of the *disappearing computer*; a palpable impediment to intuitive interaction between user and computing environment. Placing continua, the reality-virtuality and the *Weiser's continuum* at right-angles opens a 2D space shown in Figure 2 in which different application domains represent areas in this space. This thesis concentrates on the third quadrant (lower left corner) and describes applications that enhance the real environment with registered virtual information overlays, In particular for geographic information systems and mobile mapping services.

There is increasing interest in linking augmented reality with cartography and the geospatial domain. For example, Schmalstieg and Reitmayr describe how to employ AR as a medium for cartography (Schmalstieg and Reitmayr 2006). While ubiquitous computing aims at computers becoming embedded and invisible in the environment, AR focuses on adding information to the reality. With the advent of mobile computers, the confluence of these paradigms is happening in the form of mobile AR.

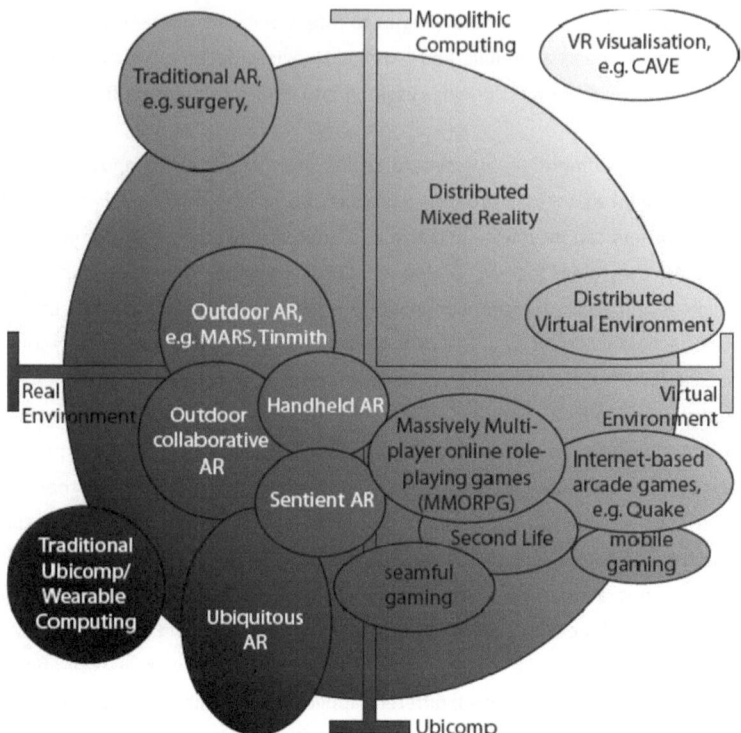

Figure 2: Milgram-Weiser diagram of (Newman et al., 2007). The diagram shows the relation between the reality-virtuality continuum and the Weiser's continuum.

1.3   Problem statement

Typically, the development process of an AR application is complex, and that is even more true for mobile AR applications. In order to build high-quality mobile AR setups, it is necessary to assemble various hardware components and sensors. Only high-quality sensors allow for accurate tracking of the user and thus a proper 3D registration of the virtual content on the environment.

Moreover, currently no standardized data format for AR content and models exits nor are there ways of generating 3D models efficiently. Note that there are additional requirements for AR content and models as both visualization and tracking need to be supported. For this reason, the author investigated several approaches to creating 3D models for various applications. Currently there is strong demand for AR applications for low-end hardware such as smart phones whereas there is a lack of serious AR applications in industrial settings. This is partly due to the need for higher tracking and registration quality. Therefore many constraints must be considered, including proper pose tracking and 3D modeling.

All these steps have to be done successfully in order to be able to perform experiments with AR applications. Consequently, the effort required to build a system is high, but it is vital to have a running system which can be used in experiments and evaluations.

1.4 Hypotheses

This thesis discusses the integration of 3D modeling and tracking for mobile augmented reality. It therefore poses the following hypothesis statements that are examined throughout the reminder of this document:

Every AR application needs some kinds of 3D models for visualization. When taking a closer look, it becomes evident that existing 3D models are only partly suitable for AR applications. This observation led to hypothesis 1. Moreover, up to now 3D models for AR were specifically built for the according applications. This limitation of access to 3D models represents a bottleneck for the future of the field of AR. Ways of efficiently generating 3D models for AR are needed. Hypothesis 2 focuses on this issue.

H1) AR needs special models that are different from existing models, in particular because they need to support both visualization and tracking.

H2) Global referenced 3D models for augmented reality can be created efficiently by using surveying procedures or using legacy data.

Accurate registration in 3D is in particular needed for applying AR in industrial settings. Primarily, proper registration depends on the accuracy of the 3D models and on the tracking accuracy, whereas the latter is more demanding. Hypothesis 3 was formulated to investigate if the required tracking accuracies for specific industrial tasks can be achieved.

H3) The accuracy of hybrid tracking can be sufficiently high for field worker tasks such as underground network inspection, planning and maintenance in industrial outdoor environments.

Finally, there is the question how useful an AR interface can be in contrast to conventional user interfaces. Is an AR user interface able to significantly improve workflows in industrial settings? Hypothesis 4 makes that claim.

H4) An augmented reality interface has advantages over conventional maps in outdoor industrial settings.

1.5   Contribution

Several main contributions can be elicited from the body of work presented in this thesis. Over the last six years, the author has created various mobile augmented reality systems – ranging from HMD based setups to handheld setups – and a set of applications and tools to gather experience with ubiquitous AR applications. The author focused on creating solutions for both indoor and outdoor AR. The applications require 3D models and information which is presented to the user. AR needs a strong content creation pipeline. Different accurate 3D models for indoor or outdoor use were developed to fit specific requirements, in particular the author focuses on *"world reference"* augmented reality that presents the user with local, human-scale content and virtual models within a mobile browser that augments reality. Next, from more than 20 peer-reviewed publications of the author elicited contributions are shortly summarized.

Creation of semantic geospatial 3D models for AR. Different approaches for generating geospatial 3D models have been investigated. The approaches comprise manual, semi-automatic and automatic methods. Manual methods are very accurate but it would be too time consuming to build 3D models of larger buildings or likewise larger environments (Schall, Newman, & Schmalstieg, 2005). For the modeling of indoor environments, a semi-automatic approach utilizing a mobile robot has been investigated (Newman, Fraundorfer, Schall, & Schmalstieg, 2005). In order to extend the 3D model to larger scale, automatic methods are needed. A transcoding approach is presented that contributes to the automatic generation of semantic AR models by exploiting geospatial databases (Schall, Junghanns, & Schmalstieg, 2008a). Even though the models themselves can be quite large and cover several square kilometers, the user is especially interested in the parts of the model that are in his or her vicinity or human-scale surroundings.

Hybrid pose tracking for mobile AR. Pose tracking is an integral part of every AR and VR application. The user's pose must be measured accurately, robustly and in real-time in order to achieve registered overlays in 3D. While there are many commercial tracking systems available that perfectly fulfill these requirements, these solutions typically target stationary setups. For mobile setups, high-quality hybrid tracking solutions are more difficult to develop and typically several tracking approaches need to be integrated into such hybrid tracking methods. A hybrid tracking approach for an indoor navigation application is presented as a simple example to demonstrate the need of AR models (Newman, Schall, Barakonyi, et al., 2006) (Newman, Schall, & Schmalstieg, 2006). Moreover, sophisticated hybrid tracking approaches have been developed for global pose estimation in outdoor environments (Schall et al., 2009). Using vision-based tracking, these approaches aim at overcoming the weaknesses of current sensors. For example, magnetic sensors are influenced by electromagnetic fields in the environment. A north-centered orientation tracking approach (reported in Schall, Mulloni, and Reitmayr 2010) considers the alignment of virtual content to true north. To experiment with different tracking approaches, the hardware setups including the necessary sensors needed to be built first. The author has designed and constructed various AR setups, among them the currently smallest mobile outdoor AR system capable of RTK GPS tracking, inertial and vision-based tracking.

Mobile AR applications. The developed systems have been deployed in various mobile AR applications for indoor and outdoor scenarios. A very interesting and promising area for AR to be applied in the industrial domain is described in a book chapter about the project *Vidente* (Schall, Junghanns, & Schmalstieg, 2010). Results reported in (Schall, Mendez, & Kruijff, 2008b) give evidence that AR interfaces have advantages over conventional maps in outdoor industrial settings and can fulfill the demanding requirements for this setting. The advantages of not only visualizing geospatial 3D models but also the interaction with them has been shown and discussed in (Schall, Mendez, & Schmalstieg, 2008b)

Evaluating AR applications with real users from industry. Until recently applications have typically been evaluated with test users in test environments (research labs). Bringing AR to the real users, e.g. utility companies, and making it run on their devices allows evaluation usage in natural environments and conditions. Results from qualitative evaluations with real-world users are reported in (Schall, Mendez, & Schmalstieg, 2008) and (Schall, Mendez, & Kruijff, 2008b).

Altogether, the body of work provides lessons learned and insights in research activities with the vision of building high-quality AR applications and bridging the gap between pose tracking and semantic 3D content creation.

1.6   Collaboration statement

This section provides a commented list of peer-reviewed publications in which results of this thesis have been published and on which chapters in this work have been based. This thesis builds upon work done in collaboration with other researchers. The following researchers deserve specific mention:

Istvan Barakonyi, collaborated with Ubisense tracking experiments
Fritz Fraundorfer, performed robot based surveying experiments indoors
Helmut Grabner and Michael Grabner, contributed with their boosting tracker
Sven Havemann, contributed with the procedural modeling framework
Sebastian Junghanns, provided in-depth industry knowledge and helped preparing and performing various demos
Denis Kalkofen, contributed with visualization techniques
Manfred Klopschitz, assisted in performing tracking tests
Ernst Kruijff, (co)-designed and (co)-developed the hardware setups; moreover, he contributed with valuable input for user interviews and evaluations
Franz Leberl, contributed with visions to the field of visual computing and insightful discussions
Erick Mendez, contributed with visualization techniques to the Vidente project, preparing and carrying out many live AR demos
Alessandro Mulloni, assisted in performing orientation tracking tests on phones
Joseph Newman, helped installing indoor tracking environments and performing indoor tracking experiments
Horst Bischof, supported several papers about computer vision with outstanding knowledge and engagement
Monika Ranzinger provided in-depth industry knowledge from the utility sector
Bernhard Reitinger, contributed with 3D point reconstruction work

Gerhard Reitmayr, contributed on the sensor fusion tracking approaches, coding sessions and provided insights in very valuable discussions

Dieter Schmalstieg, made the work possible and provided a stimulating environment

Andreas Schürzinger, gave user input and tested AR prototypes

Elise Taichmann, collaborated with the hybrid outdoor tracking approach

Eduardo Veas, (co)-designed and developed hardware setups

Daniel Wagner, supported the research with in-depth discussions and input over several years; along with others, he contributed with vision-based tracking software

Paul Wohlhart, contributed with work on vision-based indoor pose tracking

Stefanie Zollmann, contributed with visualization techniques for the Smart Vidente project, preparing and doing manifold live AR demos

The following papers describe mobile AR prototypes and applications which the author has built in order to implement applications for experiments and research purposes. The hardware prototypes are equipped with various sensors and are employed in both indoor and outdoor application domains.

Schall G., Reitinger B., Mendez E., Junghanns S. Schmalstieg D., "Handheld Geospatial Augmented Reality Using Urban 3D Models", Workshop on Mobile Spatial Interaction in conjunction with ACM International Conference on Human Factors in Computing Systems (**CHI 2007**), San Jose, USA, 2007.

> Position statement describing a first functional prototype of the outdoor handheld AR platform and application. Personal contribution: writing, idea and implementation of tracking and geospatial modeling, building hardware setup and outdoor testing.

Schall G., Junghanns S., Schmalstieg D., "VIDENTE - 3D Visualization of Underground Infrastructure using Handheld Augmented Reality", "**Geohydroinformatics - Integrating GIS and Water Engineering**", Francis & Taylor, ISBN: 9781420051209.

> Book chapter can be regarded as an integrated description of the Vidente project. Personal contribution: writing, idea and implementation of tracking and geospatial modeling.

Schall G., Schmalstieg D., Leberl F., "Einsatz von Mixed Reality in der mobilen Leitungs-auskunft", 15. **Internationale geodätische Woche**, Obergurgl, 8-14 February 2009.

Personal contribution: writing of the resume and summary of the application of AR for industrial purposes.

Schall G., "Handheld Augmented Reality in Civil Engineering", 4th conference on computer image processing and its application in Slovenia 2009 (**ROSUS'09**), Maribor, Slovenia, 19 March 2009.

Personal contribution: writing of discussion of AR in civil engineering.

Schall G., Mendez E., Kruijff E., Veas E., Junghanns S., Reitinger B., Schmalstieg D., "Handheld Augmented Reality for Underground Infrastructure Visualization" In **Personal and Ubiquitous Computing**, Special Issue on Mobile Spatial Interaction, **Springer**, 2008.

Springer journal paper presents as a main contribution expert interviews and evaluations of the handheld AR platform as well as the Vidente application. Personal contribution: writing, idea and implementation of tracking and modeling approach, hardware (co)-development, performing evaluations and user tests.

Schall G., Mendez E., Schmalstieg D., "Virtual Redlining for Civil Engineering in Real Environments", In proceedings International Symposium on Mixed and Augmented Reality 2008 (**ISMAR'08**), Cambridge, UK, 15-18 September 2008.

This paper presents an important interaction (annotation) feature called redlining. Personal contribution: writing, concept, generating geospatial models, implementation tracking approach and performing user evaluations.

Junghanns S., Schall G., Schmalstieg D., "Employing location-aware handheld augmented reality to assist utilities field personnel", Proceedings of the 5th International Symposium on LBS & TeleCartography (**LBS'08**), extended abstracts volume, Salzburg, Austria 26-28 November 2008.

This article largely deals with handling the data on the geospatial database side. Additionally a showcase was presented at LBS'08. Personal contribution: writing, concept.

Schall G., Mendez E., Junghanns S. Schmalstieg D., "Urban 3D Models: What's underneath? Handheld Augmented Reality for Subsurface Infrastructure Visualization", 9th International Conference on Ubiquitous Computing (**UbiComp '07**), Innsbruck, Austria, 2007.

This paper and demo is about outdoor AR. Personal contribution: writing, idea, generating geospatial models, conducting guided interviews with conference participants at Ubicomp.

Schall G., Mendez E., Schmalstieg D., "Virtual Redlining in Civil Engineering using the Handheld Augmented Reality Device Vesp´R". Int. Symposium on Mixed and Augmented Reality 2008 (**ISMAR 2008**), Cambridge, UK, 15.-18. September 2008.

Personal contribution: writing, preparing live demo of the AR application shown to attendees.

Since pose estimation is a key element for augmented reality, several new hybrid tracking approaches have been implemented and investigated by the author. The first three articles deal with the deployment of fiducials in indoor environments and show experiments in combination with other sensors allowing for hybrid tracking on backpack/handheld AR platforms.

Schall G., J. Newman, D. Schmalstieg, "Rapid and Accurate Deployment of Fiducial Markers for Augmented Reality", In proceedings of the 10th **Computer Vision Winter Workshop**, Zell an der Pram, Upper Austria, 2005.

This workshop paper describes how to prepare indoor environments suitable for AR. Personal contribution: writing, idea, surveying indoor environment and preparing and performing tracking tests

Newman J., F. Fraundorfer, G. Schall, D. Schmalstieg, "Construction and Maintenance of Augmented Reality Environments using a Mixture of Autonomous and Manual Surveying Techniques", In proceedings of the 7th conference on **Optical 3-D Measurement Techniques**, Vienna, 2005.

Personal contribution: building marker based tracking infrastructure and performing experiments.

Newman J., G. Schall, I. Barakonyi, A. Schürzinger, D. Schmalstieg, "Wide-Area Tracking Tools for Augmented Reality", In proceedings of the 4th International Conference on **Pervasive Computing´06**, Dublin, 2006.

This workshop paper describes a hybrid tracking approach that has been demonstrated live at the conference. Contribution: idea and implementation of tracking approach, preparing on-site tracking infrastructure and live demos.

Schall G., Grabner H., Grabner M., Wohlhart P., Schmalstieg D., Bischof H., "3D Tracking Using On-line Keypoint Learning for Mobile Augmented Reality", Workshop on Visual Localization for Mobile Platforms in conjunction with IEEE Conference on Computer Vision and Pattern Recognition (**CVPR 2008**), Anchorage, Alaska, USA, 2008.

This article is a publication at a CVPR workshop and deals with on-line tracking of an unknown target solely using the camera as sensor. Personal contribution: writing, idea, concept, implementing mobile AR setup, generating geospatial models.

The following four publications describe the integration of visual tracking approaches for global pose estimation. The approaches facilitate hybrid sensors for a multi sensor fusion method.

Schall G., D. Wagner, G. Reitmayr, E, Taichmann, M. Wieser, D. Schmalstieg, B. Hofmann-Wellenhof, "Global Pose Estimation using Multi-Sensor Fusion for Outdoor Augmented Reality", In proceedings of the International Symposium on Mixed and Augmented Reality 2009 (**ISMAR´09**), Orlando, USA, 19-24 October 2009.

This full paper presents as main key contribution a sophisticated outdoor AR hardware setup and a hybrid tracking approach eliminating the electromagnetic influences of the environment. Personal contribution: writing, idea and the full implementation of sensor fusion approach and hardware setup, generating geospatial models, preparing test environment, performing experiments.

Reitmayr, G. Langlotz, T. Wagner, D. Mulloni, A. Schall, G. Schmalstieg, D. Qi, Pan, Simultaneous Localization and Mapping for Augmented Reality, Ubiquitous Virtual Reality (**ISUVR 2010**), 2010 International Symposium, p. 5-8.

This paper presents an approach for simultaneous localization and mapping. Personal contribution: idea and full implementation of sensor fusion of inertial and vision-based tracking.

Schall G., A. Mulloni, G. Reitmayr, "North-Centered Orientation Tracking for Mobile Phones", In proceedings of the International Symposium on Mixed and Augmented Reality 2010 (**ISMAR´10**), Seoul, South Korea, 13-16 October 2010.

> This poster presents an approach on how to better register virtual content on the environment. Personal contribution: writing, idea and implementation of hardware setup, generating geospatial models, preparing test site, performing experiments.

The following three publications (one conference publication and two CG&A journal publications) deal with generating and managing 3D models for AR.

Newman J., G. Schall, D. Schmalstieg, "Modeling and Handling Seams in Wide-Area Sensor Networks" In proceedings of the 10th IEEE International Symposium on Wearable Computers (**ISWC´06**), Montreaux, Switzerland, 2006.

> Short paper presented at ISWC on indoor tracking and modeling. Personal contribution: idea and implementation, hardware and tracking setup, generating geospatial models, preparing test site, performing experiments.

Schmalstieg D., Schall G., Reitmayr G., Newman J., Wagner D., Ledermann F., Barakonyi I., "Managing Complex Augmented Reality Models" In IEEE Computer Graphics and Applications (**CG&A 2007**), Special Issue on 3D Documents, no. 4, July/August 2007.

> Journal article presenting a framework for managing complex AR models. Personal contribution: generating geospatial indoor models, preparing tracking infrastructure, implementation indoor navigation application.

Mendez E., Schall G., Havemann S., Junghanns S., Schmalstieg D., "Generating 3D Models of Subsurface Infrastructure through Transcoding of Geo-Databases" In IEEE Computer Graphics and Applications (**CG&A 2008**), Special Issue on Procedural Methods for Urban Modeling, no. 3, May/June 2008.

> This journal article describes an approach to how 3D models can be generated and employed while keeping their semantic meta data. Personal contribution: idea and implementation of transcoding pipeline, generating of semantic geospatial models, performing transcoding experiments.

The following two articles (conference paper and workshop paper at GISCIENCE 2008) describe a transcoding pipeline for generating wide-area 3D models of the urban underground using geospatial data sources.

Schall G., Schmalstieg D., "Interactive Urban Models Generated from Context-Preserving Transcoding of Real-World Data", Proceedings of the 5th International Conference on GIScience (**GISCIENCE 2008**), abstracts volume, Park City, Utah, USA, 23-26 September 2008.

    Personal contribution: Design and implementation of the transcoding pipeline for generating 3D models for AR.

Schall G., Junghanns S., Schmalstieg D., "The Transcoding Pipeline: Automatic Generation of 3D Models from Geospatial Data Sources", Workshop on Trends in Pervasive and Ubiquitous Geotechnology and Geoinformation in conjunction with the 5th International Conference on GIScience (**GISCIENCE 2008**), Park City, Utah, USA, 23-26 September 2008.

    Personal contribution: Design and implementation of the transcoding pipeline for generating 3D models for AR.

1.7   Organization

Following the introduction of the thesis, Chapter 2 provides an overview of relevant background information. Chapter 3 discusses the major requirements for building high-quality mobile AR applications. The approaches and solutions in the following chapters are directly derived from an analysis of the requirements of applying AR techniques within ubiquitous computing applications.

    Chapter 4 discusses several techniques for generating geospatial models from different sources. The author reviews recent developments in the geographic information systems community, and how they can be used by mobile AR systems. The overall focus was on achieving a high degree of automatism leading to a strong content creation pipeline. A common global reference frame was used for the generation of indoor building models (Newman, Schall, & Schmalstieg, 2006) as well as for larger outdoor models spanning dozens of buildings as well as underground infrastructure (Schall & Schmalstieg, 2008), (Schall, Junghanns, & Schmalstieg, 2008a), (Mendez et al., 2008).

Chapter 5 summarizes the authors work on mobile outdoor AR and reports on hardware platforms for indoor use (Newman, Schall, & Schmalstieg, 2006) and for outdoor use (Schall, Mendez, & Kruijff, 2008b). Moreover, hybrid outdoor tracking solutions are discussed including user interface considerations and test applications. How can high tracking accuracy, robustness and stability be realized? This question is handled in Chapter 6. A simple approach for 6DoF hybrid indoor tracking was presented in (Schall, Newman, & Schmalstieg, 2005), and with using natural feature tracking (Schall et al., 2008a). Material for describing a hybrid tracking approach for 6DoF tracking in outdoor environments is described in (Schall et al., 2009) and (Schall, Mulloni, & Reitmayr, 2010).

Building on the described approaches for modeling and tracking, an outdoor AR application is presented in Chapter 7 (Schall, Mendez, & Schmalstieg, 2008b), (Schall, Junghanns, & Schmalstieg, 2010). Finally, Chapter 8 summarizes the work and presents conclusions.

2.   Background and related work

2.1   Augmented reality displays

The augmented reality prototype built by Sutherland in 1968 used a head-mounted display as illustrated in Figure 3. These head-mounted displays required expensive equipment, limiting augmented reality to labs and research institutions. Indeed this is where augmented reality stayed until the rise of mobile computing platforms which researchers equipped with external sensors for tracking. Recent mobile computers such as smart phones are already equipped with sensors, namely GPS, digital compasses and cameras, enabling the realization of augmented reality.

Generally, AR displays can be split into head-mounted displays (HMD), handheld displays and projection displays, the latter being stationary and potentially able to accommodate multiple users.

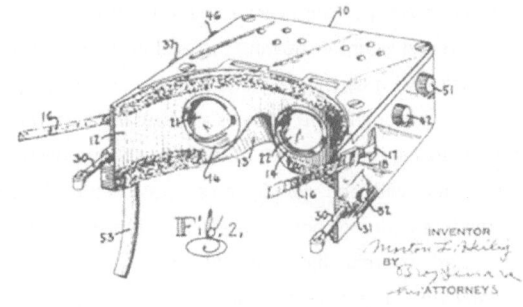

Figure 3: The first head-mounted display by Ivan Sutherland.

Also, for image generation and merging with the real world, two approaches can be distinguished as described by (Schmalstieg and Reitmayr 2006): optical see-through systems, which allow the user to see through the display onto the real world, and video see-through systems, which use video cameras to capture an image of the real world and provide the user with an augmented video image of the environment. As a result, five major classes of AR can be distinguished by their display type and their merging approach: optical see-through HMD AR, video see-through HMD AR, handheld display AR, projection-based AR with video augmentation and projection-based AR with physical surface augmentation.

Projection-based AR with video augmentation uses video projectors to display the image of an external video camera augmented with computer graphics on the screen whereas projection-based AR with physical surface augmentation projects light onto arbitrarily shaped real-world objects. It uses the real-world objects as the projection surface for the virtual environments. Ordinary surfaces have varying reflectance, color and form. Limitations of mobile devices, such as low resolution and small field of view, focus constraints, and ergonomic issues can be overcome in many cases by the utilization of projection technology. Applications that do not require mobility can benefit from efficient spatial augmentations. The focus of the work is on mobile AR systems.

2.2   Mobile augmented reality

A mobile AR system can present three-dimensional information superimposed on a roaming user's view of a task location. A decade ago, HMDs were widely built and employed by research groups. More recently, the focus has shifted towards smaller handheld devices, as depicted in Figure 4 showing the evolution of mobile AR systems. Head-mounted displays are usually worn by the user and provide two image-generating devices, one for each eye. Optical see-through HMD AR uses a transparent HMD to blend together virtual and real content. Video see-through HMD AR uses an opaque HMD to display merged video of the virtual environment with and view from cameras on the HMD. By overlaying the video images with the rendered content before displaying both to the user, virtual objects can appear opaque and occlude the real objects behind them.

Early work on mobile AR, such as the Touring Machine from (Feiner, MacIntyre, Höllerer, & Webster, 1997) used backpacks with laptop computers and HMDs (see Figure 5). (Höllerer, Feiner, Terauchi, Rashid, & Hallaway, 1999) built a series of mobile AR systems (MARS) prototypes, starting with extensions to the Touring Machine.

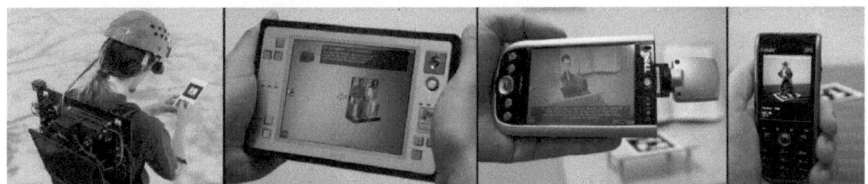

Figure 4: Evolution of mobile AR systems. Hardware setups range from backpack systems to handheld computers.

Similar augmented reality prototypes have been built by Piekarsky and Thomas in form of the Tinmith system (Piekarski & Thomas, 2001). The Tinmith-Metro application is the main application that demonstrates the capture and creation of 3D geometry outdoors in real-time, leveraging the user's physical presence in the world. Furthermore, (Reitmayr & Schmalstieg, 2004) have shown a collaborative augmented reality application for outdoor navigation and information browsing. However these systems are rather cumbersome for mobile applications deployed over longer working periods. Research focused on smaller, more convenient prototypes.

With the advent of handheld devices featuring cameras the video see-through metaphor has been widely adopted for AR systems providing augmented or "X-ray vision" views to the user. Consequently, handheld AR displays also use the video-see-through approach (Schmalstieg & Reitmayr, 2006). However they can be built from tablet PCs, Ultra Mobile PCs, or even mobile phones and devices which are highly available, and have good technical and ergonomic acceptance.

Figure 5: Mobile AR systems. (Left) Situated Documentaries. User with MARS prototype based on backpack including differential GPS. (Right) Tinmith prototype from University of South Australia.

Figure 6: Going out. (Left) A user operating a handheld augmented reality unit tracked in an urban environment. (Middle) Live shot showing the unit tracking a building. (Right) Screenshot from a pose close to the left images with overlaid building outline.

Recently, handheld AR displays became popular and can be potentially used in ubiquitous computing. This alternative and more ergonomic approach based on a handheld computer was originally conceived by (Fitzmaurice & Buxton, 1994), and later refined into a see-through AR device by (Rekimoto, 2001). UMPCs are basically small mobile PCs running standard operation systems. This has started a strong trend towards handheld AR (Wagner, Pintaric, Ledermann, & Schmalstieg, 2005). For example, Kruijff and Veas designed a two-handed shell around an Ultra Mobile PC (Kruijff & Veas, 2007).

Moreover, Reitmayr and Drummond demonstrated a vision-based tracking approach on an UMPC as depicted in Figure 6 (Reitmayr & Drummond, 2006). Today, already smart phones are fully featured high-end cell phones with GPS, camera, inertial sensors and a GHz processor, so that applications for data processing and connectivity can be installed on them. These sensors reflect the state of the sensors used for early backpack AR prototypes. As the processing capability of smart phones is improving, this enables a new class of AR applica-tions which use the camera also for vision-based tracking. Notable examples are from (Wagner, Reitmayr, Mulloni, Drummond, & Schmalstieg, 2008) utilizing them as final mobile AR displays (see Figure 7). In 2009 a promising approach was implemented within the Wikitude project (Breuss-Schneeweis, 2009), basically implementing a mobile AR travel guide with AR functionality based on user-generated Web2.0 Wikipedia or Panoramio content. The user sees an annotated landscape, mountain names or landmark descriptions in an aug-mented reality camera view. The problem with such approaches is that tracking solely relies on GPS and magnetometer which is leading to a poor registration. Therefore, latest research on smart phones focuses on vision-based tracking of natural features to overcome these drawbacks. The approaches should allow tracking the user in unprepared and unconstrained environments. Also Rohs used smart phones for markerless tracking of magic lenses on paper maps in real-time (Rohs, Schöning, Krueuger, & Hecht, 2007).

Consider that mobile AR can be realized on a variety of hardware platforms depending on the user group, requirements and the specific tasks. Among the recurring themes of AR research are world-registered (augmentable) annotations. Mobile AR is specifically suited for mobile spatial interaction (Fröhlich, 2009). Experience showed that the mobile user is very interested in interacting with the AR models. For example, (Thomas & Piekarski, 2002) experimented with spatial interaction with 3D models (see Figure 5 (right)). Lately, (Wither, DiVerdi, & Höllerer, 2009) investigated annotations for augmented reality. Typically, mobile spatial interaction is performed on superimposed geospatial 3D models or new annotations are spatially fixed to the real-world view. (Paelke & Brenner, 2007) investigate interaction with spatial data considering which tasks are relevant and consider the scope of the interaction. Interaction tasks include identification of objects, information about objects, localization of objects, user guidance, navigation, spatial selection, spatial positioning, and data collection.

Such procedures enable users with the capability of for example on-site documentation, interactive placement or correction of information. For example, in industrial settings this would be useful for field workers of utility companies aiming to locate particular items of the underground infrastructure. More generally, this procedure is also useful for city tourists using their smart phone for leaving annotations in the space. Next, the author examines what kind of data and data sources can be utilized for building geospatial 3D models that can be used for AR.

Figure 7: A user operating a smart phone using an AR application for visualizing labels registered on the environment.

2.3 Geospatial models

Mobile augmented reality requires geospatial data to present world-registered overlays. AR has a strong demand for a content generation pipeline. Currently, the process of generating 3D models for AR is not fully investigated. A leak of models for AR can be a bottle-neck for the future growth of AR applications. It seems reasonable to exploit already existing data stored in databases. Furthermore, large productive geospatial databases are the result of hundreds of person years of surveying effort. For example, a procedure of turning raw geospatial data, which are mostly 2D, into 3D models suitable for standard rendering engines could help providing manifold models for AR.

Geospatial data – also known as geographic data – refers to a particular kind of data, which is spatially referenced to the surface of the earth. The data is typically organized in geodatabases, which in turn are implemented and managed using geospatially enabled database management systems (GeoDBMS). (Schmalstieg et al., 2007) proposed a pipeline for managing AR models along the lines of a conventional information processing pipeline, which has as its main stages acquisition, storage, delivery, and use of the data. This organization separates creation and use of AR data into distinct phases. The long-term goal of mobile AR is to let users move unconstrained throughout a wide-area, and to continuously provide assistance for a wide variety of tasks. This requires coverage of the whole area and all the possible contained tasks in the underlying AR model. Scaling AR models to such wide-area-modeling coverage is only practical by leveraging legacy databases, such as existing digital maps. Manual methods for the creation of 3D models for AR are typically time consuming (see Figure 8).

The most common way to interact with geospatial data held in geodatabases is by means of a geographic information system (GIS). According to (Bruenig & Zlatanova, 2006) a GIS is a powerful set of tools for collecting, storing, retrieving at will, transforming and displaying spatial data from the real world. Sophisticated tools for spatial analysis permit to generate information relevant to decision making from the data held in underlying geodatabases. GIS can also connect to data sources other than geodatabases such as satellite imagery provided in a specific file format or Web services delivering imagery or collections of features. GIS play a major role in the context of spatial asset management (utilities and telecommunication), mapping and cadastral surveying, navigation and location- based services, planning and spatial business analysis. GIS have been available since the late 1970s by then running as monolithic stand-alone systems. In the 1990s GIS shifted towards desktop-based but still stand-alone applications. Recent developments show an increasing integration of GIS into enterprise-wide solutions where GIS communicates directly with other systems by means of Web services. In resent years a trend towards mobile GIS and 3D GIS is observable. Exploiting data stored in GIS allows for rapid and on the fly generation of three-dimensional representations of the data. Moreover, data sources such as CAD construction drawings present a huge reservoir for semantic geospatial models that can be extracted and applied in AR applications.

The vision can also be inferred from the trends in GIS research mentioned by geoscientists (Huisman & Forer, 2008):

– The increasing availability and use of shared derivative data artifacts;
– The increasing demand for temporal and dynamic functionality in geo-information;
– The increasing seeking for representations of objects true to their nature.

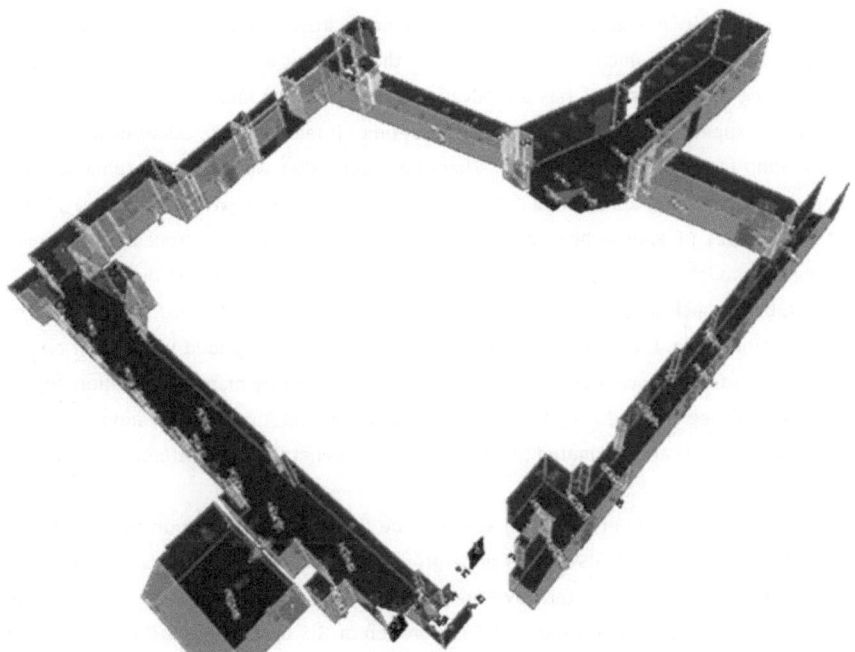

Figure 8: Manually generated 3D model of a building floor. The model includes a corridor and few dozens rooms.

This raises various issues with regard to the deployment of geospatial data. Specifically, the last trend highlights the importance of AR for providing realistic 3D visualizations for mobile GIS applications. Three-dimensional representation and visualization of geospatial environments are employed in an increasing number of applications, such as urban planning, urban marketing and emergency tasks. Existing urban 3D models can differ for example in data formats, level-of-detail (LoD) or type of data they are based on. Figure 9 shows a detailed 3D model of a city. Typically for outdoor visualizations in AR very simple models, such as building wireframes, are used.

Users often expect reliable data representations, so strict dependence on real-world measurements is necessary. Consequently, data formats based on standard GML ("Geography Markup Language | OGC®", 2010) are suitable. There are derivatives of GML, such as CityGML (Kolbe, Gröger, & Pluemer, 2008), which is a specialization of the GML language for 3D visualization 3D city models requiring a special browser. Instead, also a standard scene-graph structure can be used which enables to preserve the semantic data from the geo-database in the resulting 3D models. This has the advantage that semantic information can be used to change the appearance of the 3D model in real-time. (Mendez et al., 2008) describes such visualization techniques in more detail. There has been other work on forwarding database information to scene-graphs with a database, for example X-VRML (Walczak & Cellary, 2003), but these types of approaches generally do not involve on the fly procedural modeling. Storing the model data in a geospatial database provides the user with all the advantages of geo-databases, such as data access control, data loss prevention etc. (Bruenig & Zlatanova, 2006). Furthermore, the pipeline approach creates considerable added value from an economic point of view since a geospatial database can be used by many visualization applications (Schmalstieg et al., 2007). In addition, redundancy and inconsistency among spatially overlapping models are eliminated since all models are generated with reference to the most up-to-date data.

The work in (Roberts et al., 2002) seems to be the only AR application that is explicitly concerned with exploiting GIS data of underground infrastructure. (Paelke & Brenner, 2007) presented an AR device for interactive on-site visualization of geospatial models. Handheld devices exist that have been used in the exploration of GIS data. These include ARVino, exploring viticulture data (King, Piekarski, & Thomas, 2005), and simple landscape visualization system (Priestnall & Polmear, 2006). For most users the pure visualization of geospatial 3D models can be seen as the basic use case.

But even more useful is the ability to interact with the geospatial model and annotate a model.

LoD0   LoD1   LoD2

LoD3   LoD4

Figure 9: 3D city model and building models in various detail based on CityGML.

Note that AR models need to fulfill requirements for both visualization and tracking, thus including visual and non-visual information. Widespread future adoption of augmented reality technology will rely on a broadly accessible standard for authoring and distributing content with, at a minimum, the flexibility and interactivity provided by current Web authoring technologies.

For example, (Hill, MacIntyre, & Gandy, 2010) introduced KHARMA, an open architecture based on KML for geospatial marker and relative referencing combined with standard browser supported HTML5 and JavaScript technologies for content development and delivery. However all the approaches mentioned above assume that the data for generating 3D models are available at the required level of detail and accuracy.

Another important question is: "How can a user get access to geospatial models in a ubiquitous environment"? The real-time delivery of maps over the Internet to mobile users is still in its infancy. Increasing interactivity requires that the Web-based infrastructures enable the delivery of both 2D and 3D geospatial data to the mobile user. In this context, multiple representations of geospatial objects linked the ones with the others are desired to allow navigation at different levels of detail, representation or scales. Moreover, the representations of digitalized or independently captured data need to be consistent. Additionally, online processes, also called Web services, need to be available to enable the real-time delivery, analysis, modification, derivation and interaction with the different levels of scale and detail of the geospatial data. Current geospatial Web services are very often limited to those speci-fied by the Open Geospatial Consortium (OGC) (Welcome to the OGC Website | OGC, 2010) and standardized by ISO, namely the Web Map Service (de La Beaujardiere, 2004) (service for the online delivery of 2D maps), Web Feature Service (Vretanos, 2002) and Web Coverage Service (Whiteside & J. Evans, 2006) (services for the online delivery of respectively geospa-tial vector and raster data). However according to (Badard, 2006), if these services constitute the essential building blocks for the design of distributed and interoperable infrastructures for the delivery and access to geospatial data, no processing, such as online analysis or crea-tion of new information is possible. To overcome these shortcomings various geospatial ser-vice oriented architectures were investigated by (Badard, 2006).

On demand Web services for map delivery or services such as Google Earth provide maps of cities to mobile users. In addition, Internet GIS applications in planning and resource man-agement have become more widespread in recent years. This allows for nomadic access of GIS services anyplace and anytime via the internet by using a simple web browser. Already a growing number of companies from various sectors, such as the utility or transportation sec-tor, rely on Web applications to provide their data to construction companies or customers. In this context, Internet GIS enables mobile field workers to consult the mobile GIS at the inspection site. For example, the Austrian utility company Innsbrucker Kommunalbetriebe provides a Web interface where registered users can mark the target area on the map by drawing a polygon around the area of which they want to extract information about buried assets, such as sewer pipes, electricity or water lines. AR as a novel user interface promises to go one step further and allows viewing geospatial content in relation to the real world on-site by overlaying the virtual information over the video footage. One essential question is how to generate such geospatial content or models. Here, the Web can serve as an important pool of geospatial data.

2.4   Pose tracking

Any augmented reality application relies on some kind of tracking the user's or display's pose in order to register its content in respect to the real world. This means, determin-ing position and orientation of an object is often referred to as six-degree-of-freedom (6DoF) tracking, for the six parameters sensed: position in x, y, and z, and orientation in yaw, pitch, and roll angles. 6DoF pose tracking must run in real-time, typically requiring solutions that estimate poses in less than 50 milliseconds. Furthermore it must be ro-bust under many conditions. In case tracking is lost, the system must be able to recover quickly. Much work in mobile AR has focused on wide-area tracking. Most commercial solutions such as optical or infrared trackers cover only a limited work area, so researchers have aimed at using for example GPS (Höllerer, Feiner, Terauchi, Rashid, & Hallaway, 1999), inertial sensors (Bachmann et al., 2002), and vision (Ribo et al., 2002) for tracking. The Bat System (Newman, Ingram, & Hopper, 2001) from AT&T allows building-wide accurate tracking of people and objects outfitted with badges that are tracked by a 3D ultrasonic location system, but at the cost of deploying a building-wide electronic infrastructure.

Marker tracking is often used in AR applications if limited computational resources do not permit robust markerless tracking. One of the first projects using camera-based 6DoF tracking of artificial 2D markers was ARToolKit (Kato, Billinghurst, Blanding, & May, 1999) which was released under the GPL license and therefore became enormously popular among AR researchers and enthusiasts alike. It pioneered the use of a square planar shape for pose estimation and an embedded 2D barcode pattern for distinguishing markers. Rekimoto's 2D Matrix Code (Rekimoto, 2002) used a similar approach. Since then, many similar square track-ing libraries have emerged among which the most prominent ones are ARToolKitPlus library (Wagner & Schmalstieg, 2007)(see Figure 10 (left)).

Natural feature tracking in real-time became feasible on mobile computers since recently processing power has reached a level that allows for vision-based tracking. These approaches solve the problem of polluting the user's environment with fiducial markers. Some examples are: (Bleser, Wuest, & Strieker, 2007) uses a 3D CAD model to initialize the tracking process.

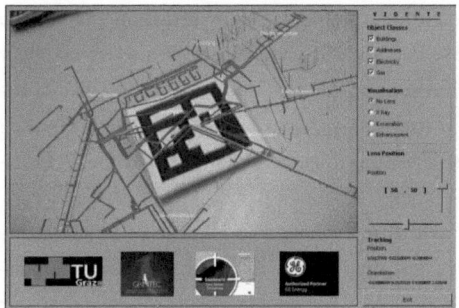

Figure 10: Table top visualization. (Left) 3D model of underground infrastructure and wireframe build-ings superimposed on a fiducial marker. (Right) PTAM (Parallel Tracking and Mapping) is a camera tracking system for augmented reality. It requires no markers, pre-made maps, known templates or inertial sensors.

The system can then extend its model of the environment automatically and even adapt to changes. (Reitmayr & Drummond, 2006) use textured 3D models of the real environment to track in urban outdoor environments (see Figure 6). A state of the art method of estimating camera pose in an unknown scene is presented by (Klein & Murray, 2009). While this has previously been attempted by adapting SLAM algorithms developed for robotic exploration, they propose a system specifically designed to track a hand-held camera in a small AR workspace (see Figure 10 (right)).

Indoor AR often relies on marker based tracking, natural feature based tracking or on installed sensor systems, such as Ubisense (Steggles & Gschwind, 2005). A user carrying a tag can be localized in 3D in the environment equipped with preinstalled sensors. In contrast outdoor AR needs to employ other techniques.

Outdoor AR systems usually rely on a combination of GPS and inertial/magnetic sensors to obtain a global 6DoF registration within the world reference frame (Azuma et al., 1999). Several outdoor AR systems already integrated differential GPS systems, although form factors, weight and usability issues had room for improvement (Höllerer, Feiner, Terauchi, Rashid, & Hallaway, 1999), (Piekarski & Thomas, 2001). While GPS provides 3D positional information, rotation is estimated from linear accelerometers (measuring the local gravity vector) and magnetic compasses (measuring the local magnetic field vector). However magnetometers suffer local and temporal magnetic influences, often leading to deviations of 10s of degrees. (Azuma, Hoff, Neely III, & Sarfaty, 2002) provide an insightful description on the performance of such sensors. Careful calibration of the magnetic sensors' scale, bias and non-orthogonal parameters, as well as influences such as hard- and soft iron effects in close proximity, can reduce the deviations between measurements and the true magnetic field vector. Calibration can be based on the assumptions of measuring the same vector under different orientations (Zhang & Gao, 2009) or measuring invariants of a setup such as the angle between the north vector and gravity vector (Hu, Liu, Wang, Hu, & Yan, 2005). However in many cases a one-time calibration is not sufficient as the errors change with time and location.

To overcome these drawbacks, visual tracking has become a corner stone of high quality AR systems providing pixel accurate overlays, but they usually rely on a model of the environment. Here, fusion with the sensors' data allows for robust performance under fast motion and tracking failures (You, Neumann, & Azuma, 2002) and provides initialization of the visual tracking component (Reitmayr & Drummond, 2007) who have shown that even highly robust natural feature tracking from IMU/vision sensor fusion is possible on a UMPC, if a detailed model of the environment is available. Since for generic outdoor environments such models may not exist, recent work of Wagner et al. has focused on an efficient orientation tracking and mapping technique (Wagner, Mulloni, Langlotz, & Schmalstieg, 2010) relative to an unknown starting point. Without knowledge of the global registration of a mapped panorama, the visualization of landmarks in an earth reference frame is not possible.

2.5   Discussion

In the literature a variety of approaches for 3DoF and 6DoF tracking have been investigated which rely on technologies such as GPS, inertial, electromagnetic, infra-red and ultra-wideband tracking. Recently there is a strong trend towards exploiting the camera as a sensor and applying visual tracking approaches in both indoor and outdoor environments. This is happening for two simple reasons. First, visual tracking approaches promise to be highly accurate, and secondly they allow tracking in unknown environments. Both advantages are highly desirable for AR applications.

Since no single approach can fully address the demands for stable, robust and accurate tracking, typically different approaches are combined. Each specific application has its own demands on the tracking solution, thus smart hybrid tracking solutions intelligently integrating appropriate approaches seem to be most promising.

The best way to predict the future is
to invent it.

Alan Curtis Kay
Computer science pioneer

3.   Requirements

AR is the first real consumer application where the lack of positioning accuracy presents a serious implementation problem. Up until now it has been about navigation or setting general location-based context where there was no good justification for precise positioning. Mobile outdoor AR requires relatively accurate position and orientation tracking to register the virtual information (e.g. a virtual urban 3D model) with the physical buildings and objects. One major challenge is achieving highly accurate tracking information.

Furthermore, the availability of geospatial models appears to be a key enabling factor for the success of applications for handheld devices. More concretely, models for AR are complex. For example, there is a need for semantic information along with geometric information to allow application-centric visualization. A further major challenge deals with the question of how to efficiently generate such models.

Classically, there are various coordinate systems involved when designing an AR application. Typically, for outdoor AR, the user's pose is given in a user reference system by GPS and inertial tracking (sensor reference frame). Moreover, different geospatial models will be represented in different coordinate frames as well (data reference frame). For better streamlining of the development process of AR applications, it is of great importance to match the involved coordinate systems to a common absolute reference frame. Generally, geospatial modeling and tracking are key building blocks for a successful mobile AR system.

3.1   Sensor reference frame

Typically, each tracking system uses its own local coordinate system or sensor reference frame. The position and orientation of the user (6DoF) need to be estimated accurately using a sensor fusion approach based on various sensors. Tracking approaches using GPS, inertial, gyroscopic and magnetic sensors, or computer vision-based methods can be applied.

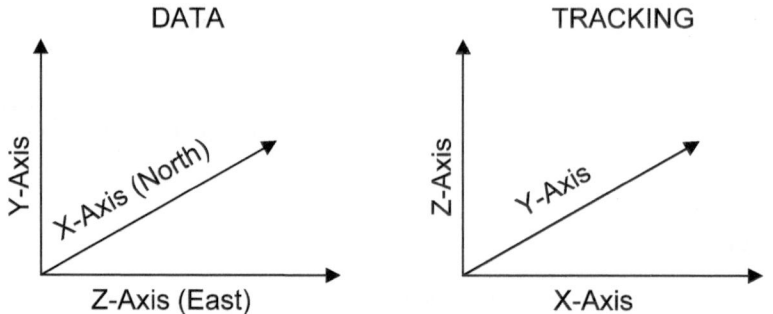

Figure 11: Coordinate systems of reference frames. (Left) Data coordinate system. (Right) Sensor coordinate system.

Today, outdoor AR applications are limited by inaccuracies in GPS/compasses in mobile devices. Figure 11 depicts the coordinate systems of the involved reference frames which are used for the applications developed by the author.

3.2   Data reference frame

Typically, geospatial models are represented in a local data reference frame and there are many different data formats using different coordinate systems. Today researchers and application developers build their own indoor and outdoor 3D models to fit the purpose of the AR application. Location information is an inherent attribute of geospatial data which is stored in various data sources. Using this data and adopting it for AR purposes would create a wealth of geospatial models to be used for AR visualizations. As a benefit, all semantic information about the geospatial data would be available for visualization and interactive purposes and for mobile spatial interaction. Currently there is a lack of solutions achieving this aim.

A common data reference frame representing geospatial models is very desirable. On a larger scale these are models of urban areas or city models; on a smaller scale these are models of buildings, rooms and interiors. Using a global location component of the geospatial data, the models could be represented in a world data model. As a very useful side effect, the existence of such a geo-referenced model would allow further enhancement of the applicability of visual tracking based on models.

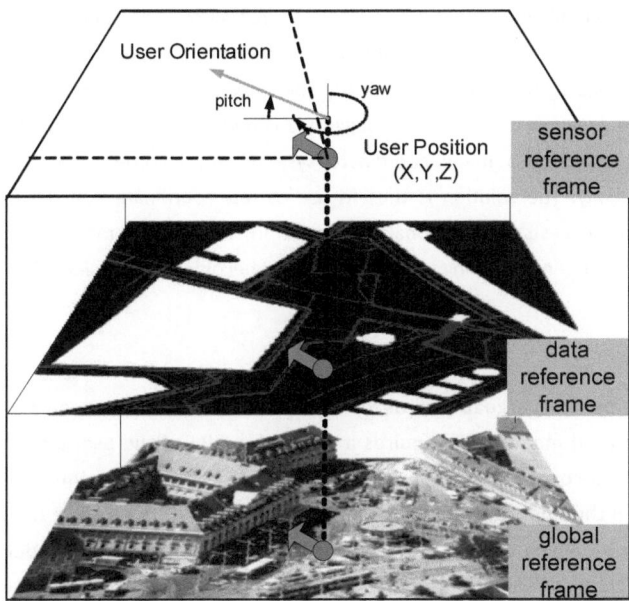

Figure 12: Reference frames. Matching of the user reference frame, map reference frame to the absolute global reference system. A major challenge is to fit the different reference systems of the base data or urban 3D models to a common absolute reference system.

3.3 Global reference frame

The author can see that problems with different reference frames would arise if different local sensor coordinate systems and data coordinate systems are involved. It is therefore a necessity to represent both geospatial models and tracking in a common reference frame to enable registration for AR on a global scale. Figure 12 illustrates the involved reference systems. Only when the sensor reference frame and the data reference frame are transformed into a common global reference frame will the tracking and data fit together.

The user's outdoor pose is given in a sensor reference frame by GPS and inertial tracking. Because it was agreed to receive GPS data in Gauss-Krüger Mercator coordinates, this coordinate system was chosen as the common global coordinate system. Consequently, the geospatial models used in the applications are also represented in Gauss-Krüger Mercator coordinate system. The GPS coordinates can also be projected to Universal Transversal

Mercator (UTM) format which allows using this coordinate format all over the world. Since geospatial base data and models use different local reference systems, a major challenge is fitting the different reference systems together. Clearly the integration of modeling and tracking is important. The previous considerations suggest a common reference frame for both modeling and tracking. If too many coordinate systems are involved, then the complexity rises exponentially. The solution is thus to use as few coordinate systems as possible. Furthermore, this suggests adding global location information to any models that are created for AR. As a consequence, sharing such models would become much easier.

3.4   Augmented reality models

Augmented reality systems have arguably some of the most difficult requirements in terms of computer generated models. AR requires a detailed model of the user's environment. The model has to consist of both visual and non-visual or semantic information. This is a main reason that models suitable for AR are more difficult to produce than models typically used for VR which solely focus on the geometric accuracy of the model. VR models and AR models have something in common: both are frequently based on measurements taken from the real environment such as architectural models. Nonetheless AR models require semantic interpretation of the environment. To illustrate this, AR uses models not only for visualization purposes but also for handling occlusions, user interface, interaction and vision-based tracking of real objects. The structure of AR models is more complex.

Besides structural complexity, the model scale is also important. The vision of AR is to allow the user to roam through unconstrained environments and be provided with useful assistance and (visual) support. Since modeling large areas manually involves a huge effort, this goal of generating large-scale AR models can only be reached by using existing legacy data for the modeling task. The basic idea is to use legacy data that is stored in geo-databases as the underlying data source upon which the AR model is generated on the fly. This has the advantages of avoiding data replication and always accessing the most up-do-date data. Moreover, bandwidth use can be optimized because only the underlying data is transferred over the network and the AR model is generated on the client side.

The AR model data helps in configuring all relevant subsystems of the AR application, including registered 3D visualization, tracking and the user interface. Classically, scene-graphs are used for visualization. A scene-graph is based on Coin3D which is a free implementation of the Open Inventor API. The scene-graph implements the application logic and can be produced on the fly. Besides the scene-graph other subsystems of the AR application also rely on dynamic model delivery. For example, the tracking system requires device configuration data and geo-referenced features for vision-based tracking.

An important question is how detailed an AR model should be. This is highly depended on the application. It does not make sense to use models with all objects presented in the highest detail. First, the user may not recognize any difference with an object represented in lower detail. Second a highly detailed model might be attained at the cost of high bandwidth usage and result in slow runtime behavior for rendering. This suggests keeping AR models compact in the sense that the representation should be kept as simple as possible but not simpler. A very interesting approach is representing models procedurally. This has the advantage of keeping the model compact by describing the objects based on simple rules. Furthermore, it allows decisions during runtime regarding which objects to render with high details and which ones not. Aside from generating geospatial models efficiently, approaches for distributed Web 2.0 technologies seem appealing. Today, AR frequently requires models to be interactive because many AR applications must allow interacting with the model.

3.5 System design considerations

As a precondition and basis for experimenting with AR applications, various hardware platforms integrating different sensors for indoor and outdoor tracking have been built. Considering the issues and requirements mentioned before, methods and approaches were investigated for efficiently building semantic 3D models for AR. The overall focus was on achieving a high degree of automatism leading to a strong content creation pipeline. A common global reference frame was used for the generation of indoor building models as well as for larger outdoor models spanning dozens of buildings und underground infrastructure networks. Furthermore, a hybrid 6DoF tracking approach for outdoor environments is presented.

Building on the described approaches for modeling and tracking, an outdoor AR application can be implemented. Different sensor reference frames and data reference frames are used. The application needs to combine the reference frames in order to achieve overlays registered in 3D. The applications benefit from the global reference frame as proposed above and illustrate how a user experience is generated by overlaying geospatial models on the real environment and providing the possibility for mobile spatial interaction. Typically the user is interested in registered information in his or her more nearby surroundings; for this reason the models need to be registered as well as possible in human-scale.

Simplicity is the ultimate
sophistication.

Leonardo Da Vinci

4.   Interactive geospatial models for augmented reality

High quality AR applications depend crucially on 3D models for visualization and interaction. The interaction between real objects and virtual content is fundamental to AR. The generation and use of 3D models for AR has come a long way from static models and offline creation to online capture during operation. Parallel to this development, 3D models already have a long history of more than 20 years in computer vision and image processing. A central question is how models can be created in such a way that they can fulfill the dual purpose of visualization and tracking. This issue will be covered for the discussed modeling approaches.

Many forms of interaction require a 3D model of the shape of the real objects involved. For example, to render a scene that combines real and virtual objects, it is vital to know how they occlude each other. For example, (Klein & Drummond, 2004) tackle this problem for a tablet-based AR system and require a pre-existing CAD model of any object in the scene if it is to occlude virtual geometry. Such approaches already assume that 3D models exist.

The following sections describe various approaches that have been implemented for the generation of 3D models. In the first approach the environment is surveyed using tachymeters and additionally employing a robot equipped with a laser rangefinder to increase the automatism in the surveying process. This approach is labor-intensive and thus limits the creation of larger models. To overcome these shortcomings, the main focus lies on the second approach that describes a transcoding pipeline generating content by taking advantage of the rich data stored in geospatial databases.

4.1   Manual surveying

This type of approach is relevant because it can be used as the "ground-truth" against which other methods can be compared. The intention is to generate an accurate 3D model that is true or correct to the real environment. This technique was used to create a 3D model for a smaller indoor environment. The resulting indoor model is structured into walls, windows, floors, doorways, implying building topology to support indoor navigation. Each of these structures need also be described by semantic information, next to its geometric information. A XML dialect called Building Augmentation Markup Language (BAUML) was used to represent the indoor model, including topological information to derive navigation hint and geo-referenced semantic information. Moreover, the model includes fiducial markers to consequently use the model for tracking purposes.

Traditional techniques using a tachymeter to survey single points were used. This technique is only applicable for smaller spaces, since the process is both time and labor consuming. In general the surveying process workflow summarized in Figure 13 consists of the following steps:

1. Firstly, preparations for surveying are undertaken. These mainly consist of making a preliminary examination of the surveying area and attaching marker templates on the walls at a maximum interval of two meters.

2. The surveying area is then divided into measurable segments. That is the area which can be measured from wherever the total station is positioned. At least one measurement is necessary for each room if there is a clear line of sight to the important surfaces and edges. Note that it is important that neighboring surveying areas overlap, ensuring that common points of correspondence (so called "pairings"), occur in both adjacent areas, acting as a bridge. A minimum of three correspondences with good geometry (i.e. neither co-linear nor too close together) are necessary in order to be able to calculate the transformation from the coordinate reference frame in one room and the coordinate reference frame in the other room. For each marker template the four corners of the square patch of the marker are measured, and the position and orientation calculated.

3. Steps 3 to 5 are performed iteratively for all parts of the surveying area. All points of the room geometry are measured using the total station, followed by the edges of portals and marker templates and at least the pairings to the neighboring room or corridor. Ids are given to all measured points, thus ensuring that each point can be uniquely identified and referenced when building the 3D model.

4. The acquired data is transferred from the total station to the PC. To assure the measurements have been conducted successfully, all points are checked.

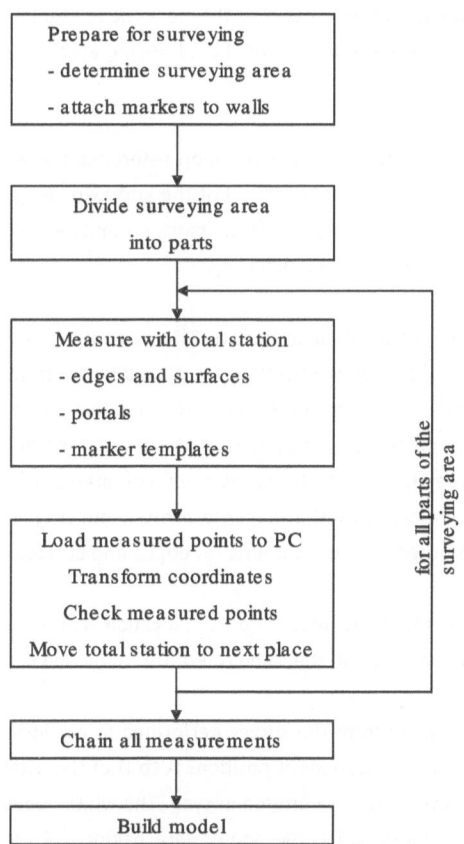

Figure 13: Manual surveying and modeling workflow.

5. The measured points are transformed from polar into Cartesian coordinate system. Scripts parse the original measurement file and convert the points into Cartesian coordinates and save them in a Matlab file format.

6. A script transforms the points from the Matlab file format into XML encoding for the representation of geometric information. It allows the building geometry (e.g. walls, floors and corridors) to be stored, as well as the positions of markers. Due to the recursive definition of the language a tree structure of spatial objects, where objects are composed of a number of smaller objects can be created.

7. Having completed these steps, the total station is moved to the next measurement position and steps 3 to 5 are now repeated in the next part of the surveying area until the entire area is surveyed.

8. Finally, all the measured points are transformed into one common reference frame. They are then chained together and merged, yielding a 3D model of the surveying area. At position A all points of measurement area A including the three pairing points P1, P2 and P3 are measured. The points in area B are determined similarly.

Figure 14 (top) shows a graph representation of the measurements. The magenta circles represent the coordinate systems of the total station where they are placed in positions A and B. The pairings, consisting of common points in the overlapping surveying area, are denoted by cyan nodes. In this case the high contrast corners of the fiducial marker are used to provide the minimal set of three, appropriately conditioned, common points required to calculate the transformation indicated by the magenta arrow from A to B. The remaining white circles represent the other measured points on edges and surfaces that are to be included in the model.

It should then be possible to transform all the points measured from location A into the coordinate system of location B. All the points are now in the same reference frame and can be chained to produce a comprehensive model.

Figure 14 (bottom) shows a subgraph obtained from the survey performed at our institute. The magenta colored circles represent the measurement positions A to D of the total station. The pairings of each part of the surveying area are printed in cyan. The white nodes represent points measured on edges and surfaces. Using the appropriate pairings all the points measured at location A are transformed into the coordinate system of location.

An approach to surveying the locations of fiducial markers has been demonstrated. The accuracy of such a model is very high and for the given model as small as a few millimeters. This was observed when performing the loop-closing of the corridor.

One needs to consider that when further extending the model, error propagation has to be considered. The general model of the environment is sufficiently accurate to act as a ground truth against which to compare SLAM experiments. But, the overall procedure for obtaining the high-quality indoor model turned out to be quite labor intensive requiring days or even weeks. The model in Figure 15 covers a corridor, 4 large rooms, doorways, windows and around 70 fiducial markers. However manual methods have two obvious drawbacks.

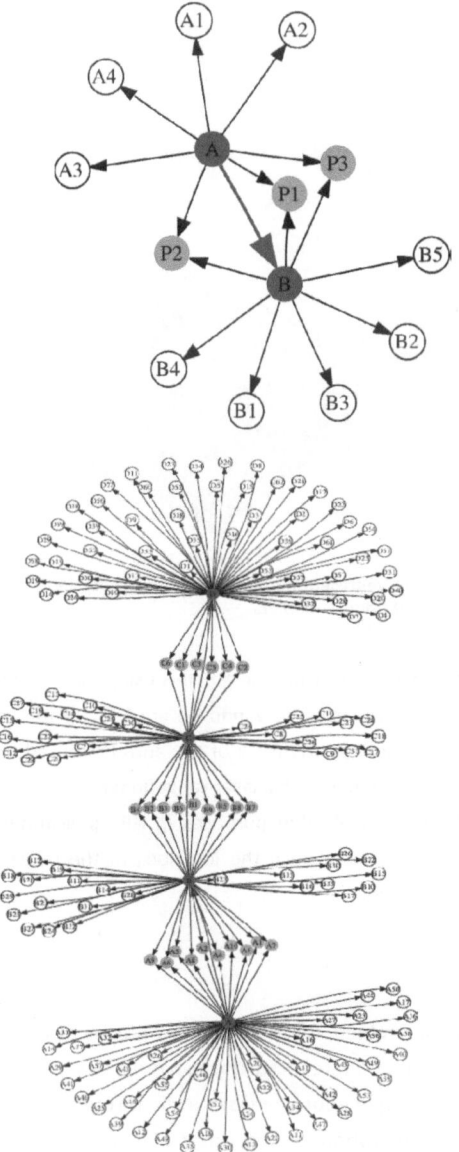

Figure 14: Spatial relationships involved in the bridging of adjacent surveying areas (top). Graph of interrelationships (bottom).

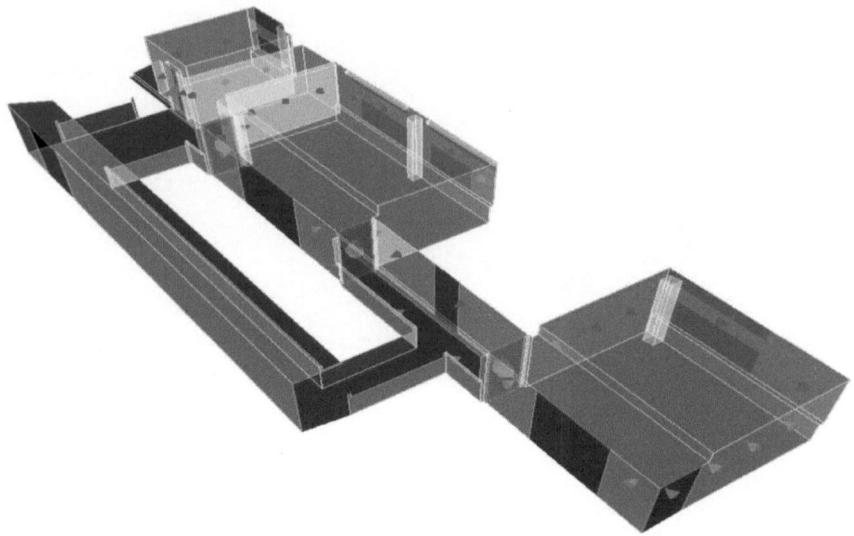

Figure 15: Indoor building model. It consists of planes representing floors, walls, ceilings, doors and windows. The cones show positions and orientations of fiducial markers.

First, they do not scale well, because the model must be constructed using many measurements. Second, certain types of building features (such as windows) are difficult to survey using these methods. For surveying larger models, early work of the author focused on semi-automated surveying using a mobile robot equipped with a laser rangefinder.

The presented model is not only useful for visualization purposes in AR applications. Moreover, it can be used for tracking the moving user since the locations of the fiducial markers are stored as additional tracking information.

4.2   Semi-automatic surveying

Most AR applications have hitherto been constrained, by the working volumes of tracking technologies, to static spaces of a few cubic meters. Furthermore, an assumption has been made that sensors are deployed homogeneously and statically throughout the area of interest, resulting in a single off-line calibration.

An approach called *Ubiquitous Tracking* attempts to automate the process of dynamically integrating arbitrary sensors in distributed sensor networks, whilst focusing on the dynamic spatial relationships in a given environment. The semantic depth of events like "person A is in room W" depends not only on the concept of a *person* that can move, but also on the concept of a *room* that cannot. Nevertheless the *room* must nevertheless be measured, and meaning assigned to these measurements. Changes in building use, and even routine maintenance, mean that that new measurements may need to be made displacing or augmenting old ones. The integrity of the spatial model depends on the complete history of measurements, and care must be taken in the assumptions that are made.

Due to the expense and limited range of current commercial trackers designed for use by the VR and AR communities, visual tracking has become very popular. At that time, when natural feature tracking was in its childhood, attempts have been made at deploying markers over a wide-area in order to extend tracking range. It is necessary to know the position and orientation of the markers as accurately as possible. Earlier surveying techniques involved the use of reflectorless total stations to survey the positions of the markers (see Chapter 4.1). This manual approach is time consuming, and presents a serious barrier to the introduction of AR to new environments. Therefore, automatic methods are necessary to speed up the process. The use of an autonomously navigating mobile robot is proposed to detect and localize the fiducial markers and build a model that can be used by existing AR systems. The mobile robot is equipped with a laser rangefinder to localize the robot as well as with a digital camera. The images taken from the camera are used to detect the fiducial markers. By fusing the 3D position of the markers with the laser based position of the robot, the absolute pose of the fiducial markers can be calculated. The effectiveness of this new approach can be assessed by comparing the model obtained from the total station with that obtained using the robot. Furthermore, a hybrid approach in which measurements from both the robot and the total station is presented.

In the previous section, techniques for rapidly and accurately surveying the locations of widely distributed markers with the theodolite-based measurement system Leica TPS 700 total station, whilst simultaneously building a model of the environment were described. A wide-area indoor tracking solution uses a set of known markers that were distributed throughout the environment. Together with a geometric model of the building that includes the location of the well-known markers the user's location can be computed as soon as a marker is tracked by the optical tracking system using ARToolkitPlus (Wagner & Schmalstieg, 2007).

Autonomous measurement of fiducial markers. The goal was to automatically create a map of deployed fiducial markers. A mobile robot capable of exploring unknown environments should detect fiducial markers and measure their position in a single coordinate system. A PeopleBot (ActivMedia) robot is equipped with a laser rangefinder (LRF) (Sick LMS 200) and a 2MP digital color camera equipped with a wide-angle lense. The data from the LRF can be used to create a floor plan of the explored area. Laser readings are taken every 5 centimeters. After a final registration of all the readings, floor plan and robot positions (including orientation) are available with high accuracy. The detection of the fiducial markers is performed using the images acquired by the digital camera. After detection, the 3D coordinates of the marker's 4 corner points are computed using stereo reconstruction. The reconstructed marker points are transformed into the overall coordinate system determined using the LRF by fusing the camera coordinate system with that of the laser pose.

Marker detection. The images captured from the camera have a resolution of 1600x1200 pixels. Color information is discarded and all subsequent steps work exclusively on grayscale images. The lense has a field of view of 90°. The camera is calibrated; interior orientation as well as lense distortion is known. In a first step, the images are resampled to compensate for the lens distortion, as the wide angle lense results in high radial distortion. To detect the markers in the images, Maximally Stable Extremal Regions (MSER) (Matas, Chum, Urban, & Pajdla, 2004) are extracted. This local detector is threshold-based and is well suited to finding the deployed markers. However this is a general approach and the detector also returns other stable image regions, which can be used as natural landmarks. The subsequent methods are applied to all detected landmarks. The classification of a detected landmark as a fiducial marker or a natural landmark is performed at the end of the mapping workflow. Figure 16 (a) shows that three landmarks have been detected by the algorithm.

Marker reconstruction. 3D reconstruction of the landmarks is done using a shape-from-motion approach (the markers are viewed from two or more different viewpoints which allows the calculation of the 3D position). Two nearby frames from the image sequence are selected and the essential matrix is calculated for the image pair. The estimation is done automatically. Harris corners are detected in both images and matched using normalized cross-correlation (see Figure 16 (b)). The essential matrix is calculated on an inlier set obtained from RANSAC using the 5-point algorithm (Nistér, 2004).

The next step is the reconstruction of the detected landmarks. As depicted in Figure 16 (b), corresponding landmarks in both images are matched using SIFT descriptors (Lowe, 2004). The matching method returns a logical matching of the landmarks as well as accurate point matches within the landmarks. For every landmark, a 3D reconstruction using the appropriate point matches is created. It is assumed that the landmarks are planar and allow a robust plane fitting in 3D. In the next step, the knowledge of the deployed fiducial markers is used to extract them from the set of all detected landmarks. This is done using basic image processing techniques. The markers in question consist of a black square surrounded by a white border. If a landmark can be identified as a fiducial marker the 4 corner points are extracted and projected it onto the plane of the landmark in 3D to obtain the 3D coordinates of the corner points. The scale factor of the metric reconstruction is determined by the knowledge of the real size of the markers (153mm x 153mm). The reconstruction is now in canonical coordinates.

The final step is to transform the single reconstructions into the overall coordinate system established from the laser scanning. For that the 3D points of the markers are rotated and translated $p_{new} = Rp + T$. R is the rotation matrix describing the orientation of the robot and thus the camera and $T = [x, y, z]^T$ is the position of the robot for the frame of the image sequence used for the reconstruction. The final result can be seen in Figure 16 (d), where the path the robot drove and the detected and reconstructed markers are drawn.

The accuracy of the results from the robot's measurements is limited because of the use of COTS (commercial off-the-shelf) components. Figure 17 depicts a superimposed image of the map of markers reconstructed from the robot (shown in Figure 16 (d)) and the map obtained by the total station including the floor plan and markers drawn as cones. In the zoomed-in area one can observe the superimposition of the markers. Figure 18 shows the floor plan of the part of the building that was measured with the laser range scanner mounted on the robot.

(a)   (b)

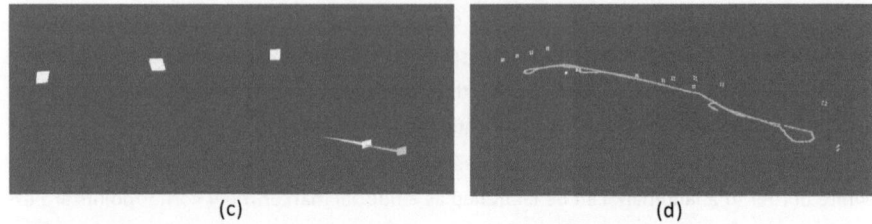

Figure 16: Semi-automatic surveying. (a) Image with 3 visible and detected markers. (b) Detected corresponding markers (c) Marker reconstruction in canonical coordinate system. (d) Map of the markers detected and reconstructed from the robot.

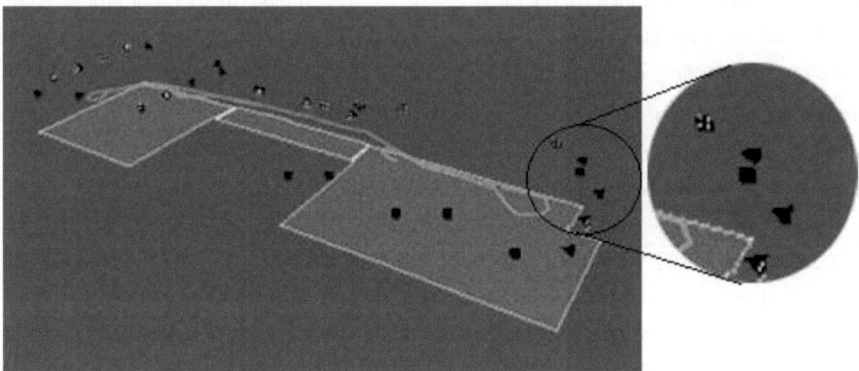

Figure 17: Superimposed map of the markers reconstructed from the robot and the map of markers measured with the total station.

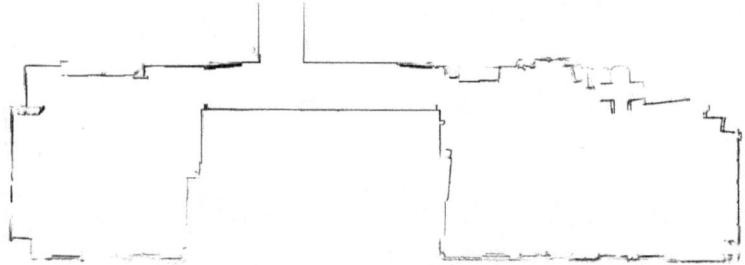

Figure 18: Two-dimensional floor plan measured by the PeopleBot robot.

The experiments showed that applying the semi-automatic approach can deliver a floor plan and the fiducial marker positions with sufficient accuracy to be use in AR applications. Note that the approaches shown are manual and semi-automatic and only applicable for smaller environments. Next, a very simple application is presented to show how the presented semantic 3D models were applied in early AR applications. The main reason to show this example is to motivate to usefulness of semantic models supporting visualization as well as tracking issues.

4.3   Example application

Although the so called "smart building" has been a dream of architects, engineers and social anthropologists alike, thus far it has proved necessary to take existing "dumb buildings" and attempt to install the necessary sensors in order to retrofit some sort of sentient behavior. In former experiments of implementing a wide-area indoor tracking solution, often fiducial markers that are part of the 3D model were used. This is an illustrative example of how additional information that is stored in the 3D model can be utilized for tracking purposes.

Wide-area tracking systems such as those based on ultrasound or UWB electromagnetic signals were promising solutions to the problem of affordable, accurate and widespread sensing of location, which is a powerful source of context. However they are limited by the reflective properties of many modern building materials. One such COTS system has been developed by Ubisense using UWB short duration pulses emitted by an active tag (Ubitag) user-worn or device-mounted. The use of both time-difference-of-arrival (TDOA) and angle-of-arrival (AOA) techniques for position calculation in the wall-mounted sensors makes it possible to locate a tag within 15 centimeters in three dimensions (Steggles & Gschwind, 2005). Such a tracking system was installed throughout the area that was manually surveyed before. Figure 19 shows the basic tracking setup.

Previous experiments with indoor navigation systems relied solely on widely distributed fiducial markers to provide a wide-area vision-based tracking capability of moderate accuracy. Newer tracking technologies, such as Ubisense's, robustly cover large areas without the visual clutter of visual markers, or the brittleness associated with natural-feature based vision trackers. This motivation lead us to explore how a wide-area tracker that can only sense position, lends itself to a hybrid approach whereby it is combined with complementary sensors to yield the pose estimates required for augmenting a user's view (Newman, Schall, Barakonyi, et al., 2006). A dynamically reconfigurable version of the OpenTracker (Reitmayr &

Schmalstieg, 2001) tracking middleware ensures that the pipes-and-filters network connecting producers and consumers of tracking information continuously adapts such that pose estimates are always available. For example, when id-based ARToolKitPlus markers (Wagner & Schmalstieg, 2007) is visible, then the pose is taken directly from the vision algorithms; however, when moving into an area where fiducials are either no longer present or are not visible due to occlusion, then the positional component of pose is taken from the Ubisense system and the orientation component of pose is taken from the inertial tracker. The sensor fusion method itself is very simple and depicted in Figure 19. The basic idea is to fuse the available sensors in a way that always the tracking system providing the highest accuracy is used.

A real ubicomp environment, its size notwithstanding, will be richly populated with objects both static and dynamic. Although originally designed for a large area requiring navigation cues, the system is still sufficiently flexible to visualize all these elements. Figure 21 shows the navigation system in action, with navigational cues, state information and current location visible using a "world in miniature" view. For the experiments the building model in Figure 15 in Chapter 4.2 has been used. The model was generated in a manual approach as described in Chapter 4.1. The used hardware setup for indoor environments is described in Chapter 5.2.

Figure 21 shows a user with a handheld AR device performing a navigation task. Full range of sensors, including fiducial markers and Ubisense wide-area tracker are utilized. Figure 21 (right) shows the user's view showing and marker, Ubisensor and a world in miniature indicating the user's position. Figure 22 (left) depicts the user's view when navigating towards the destination "corridor". The location can be determined from Ubisense wide-area tracker together with observations of fiducial markers. The direction of travel is indicated by a compass pointer shown at top right. Additionally, Figure 22 (right) shows the users view with the world in miniature and a graphical overlay of the building wireframe model registered in 3D. This AR scenario interacts with the sensors in the environment to allow for estimating the pose of a mobile user. The navigation example shows how a wide-area tracker and other sensors can adapt to meet the needs of an application in very different settings using the resources to hand. The different visualization methods for the wireframe augmentation, world in miniature and the navigation portal highlighting all reuse a single scene-graph.

Note that the very simple application of a navigation task is only feasible using the semantic information stored in the model. On the one hand, the geometric information of the 3D model is used for superimposing registered wireframes. On the other hand, the semantic information of the 3D model supports the user in the navigation task.

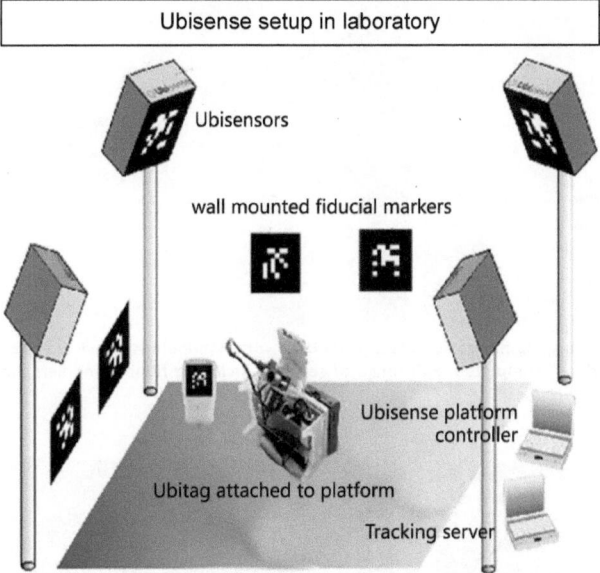

Figure 19: Indoor tracking setup. Installation with wall-mounted Ubisense sensors and a mobile platform with an Ubisense tag.

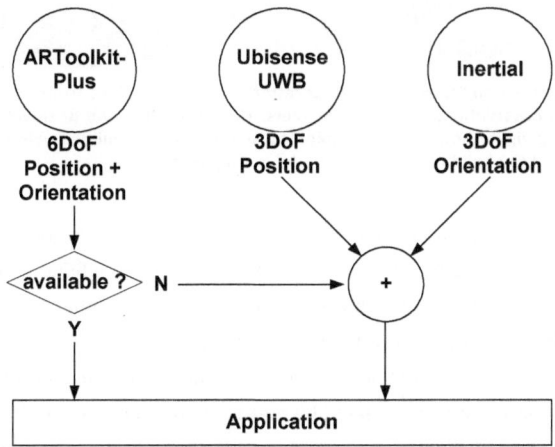

Figure 20: Hybrid tracking approach for indoor environments. If no accurate 6DoF marker tracking is performed, the system uses the complementary UWB and inertial sensors.

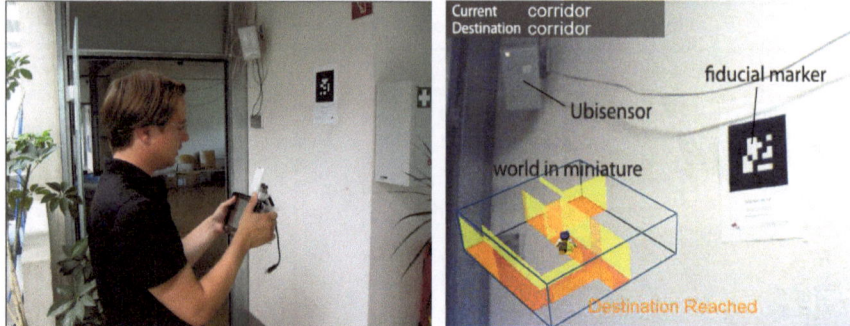

Figure 21: Handheld AR user performing navigation task. (Left) User with handheld AR device. (Right) Navigation to destination "corridor" completed. Full range of sensors, including fiducial markers and Ubisense wide-area tracker are utilized.

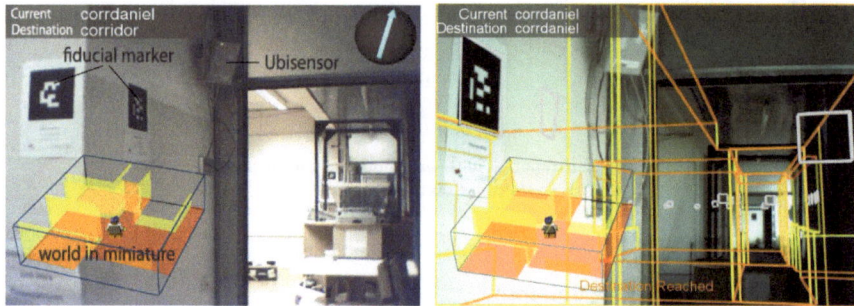

Figure 22: Navigating towards destination "corridor". (Left) Location can be determined from Ubisense wide-area tracker together with observations of fiducial markers. Necessary direction of travel indicated by compass pointer in top right shown without graphical overlay of the building wireframe model. (Right) Graphical overlay of the building wireframe model is visualized.

The user is guided, which objects are doors that he or she can walk through, in contrast to walls which are impassable objects. As already mentioned, the locations of the fiducial markers stored in the 3D model are not only for visualization purposes, but more importantly for tracking the roaming user. This example showed how early AR applications used 3D models for various purposes. In future, the AR user should be able to roam through unconstrained environments. Such large models can only be created by means of automatic generation. To achieve this vision, the following subchapter presents a promising approach for modeling large-scale outdoor models for AR applications.

4.4   Transcoding pipeline

Next, a context preserving transcoding pipeline for generation of interactive three-dimensional models is presented. These urban 3D models are based on real-world data from 2D geospatial databases and include semantic markup.

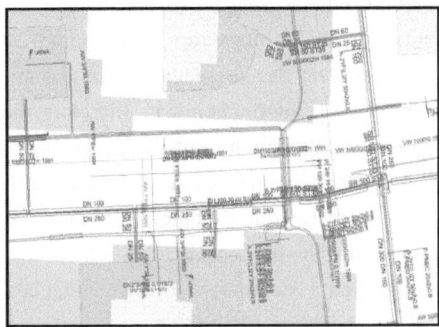 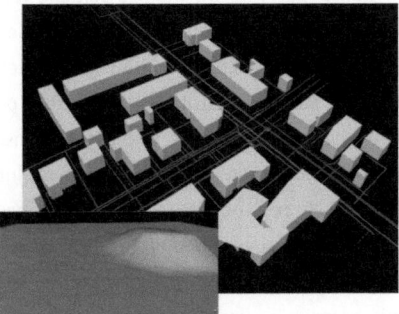

Figure 23: 2D map vs. 3D model of urban area. (Left) Cadastral plan of urban area (courtesy of Salzburg AG). (Right) Urban geospatial 3D model consisting of underground infrastructure (gas pipes, electricity lines and water mains) plus above surface features (buildings). (Data transcoded from data provided by Salzburg AG) Inset: Digital terrain model.

In contrast to the manual and semi-automatic approach, this technique promises an automatic process for generating large models.

For example, these models can then be used for interactive on-site visualization of underground infrastructure. Figure 23 shows an example of a 3D model that has been generated automatically from GIS data stored in productive geospatial databases. The geospatial model consists of underground infrastructure and extruded building footprints. Moreover, the transcoding pipeline is able to generate digital terrain models (DTM). The geospatial models are as accurate as the underlying GIS data is.

4.4.1 Transcoding process

The Oxford English dictionary ("Oxford Dictionaries Online - English Dictionary and Language Reference", 2010) gives the following explanation for the term transcoding.

> **Transcoding:** convert (language or information) from one form of coded representation to another.

The transcoding pipeline allows for a neat separation of model content and presentation. Temporary models are generated rapidly on demand from the long term, 3D models are not stored as whole but only their underlying GIS data, the rules for model generation and the styles to be applied for visualization.

To connect geospatial databases and rendering engines, the raw 2D geospatial data must be transcoded into 3D models suitable for standard rendering engines. Transcoding is not simply a one-to-one conversion from one format to another. 3D models are obtained from 2D information through procedural 3D modeling. Transcoding the geospatial database information's semantic attributes into visual primitives entails information loss. Therefore the right point in the pipeline to perform transcoding must be found. If semantic information is discarded too early, it cannot be used for interaction later in the pipeline. On the other hand, if it is discarded too late, the semantics have to be interpreted at runtime, which increases overhead and adversely affects performance. This is called the transcoding trade-off.

The implemented framework transcodes geospatial data into interactive 3D visualizations. This process is shown in Figure 24. A conventional scene-graph with semantic markup is combined with on the fly generated procedural models enhanced with an embedded stack-based scripting language.

Because these techniques are tightly integrated, the transcoding and representation methods for each object can be chosen on the basis of the available high-level semantic information. The approach also lets users define visualization styles in relation to the semantic markup, independent of actual object structures.

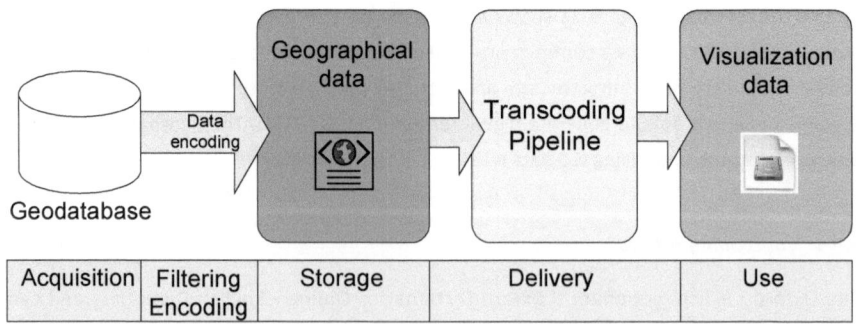

Figure 24: Transcoding pipeline. The pipeline transcodes data in GeographyML encoding to Open Inventor file format.

(Schmalstieg et al., 2007) proposed a pipeline for managing AR models along the lines of a conventional information processing pipeline, which has as its main stages acquisition, storage, delivery, and use of the data. This organization separates creation and use of AR data into distinct phases. As GeographyML encoded vector data is not suited for visualization, the transcoding step transforms the data into a format allowing efficient visualization.

GeographyML encoded features describe subsurface infrastructure objects and some above-surface objects like buildings or street-level objects such as cappings or trees.

Features consist of property attributes and one or more geometry attributes describing the actual 2D coordinates (see Appendix 9.3 in Figure 86). A separate configuration file controls which features and attributes are retained in the following transcoding step. The result of the transcoding is a scene-graph description with per-feature grouping of shape objects and semantic attributes. For details on scene-graph traversal itself, refer to (Strauss & Carey, 1992). The shape objects either refer to embedded GenerativeML scripts (Havemann & Fellner, 2004) or for static non-procedural pre-modeled shapes to Coin3D classes (see http://www.coin3d.org), a free implementation of the Open Inventor API. The Generative Modeling Language (GML, called GenerativeML in this thesis) is a simple stack-based scripting language for creating parametric 3D models (also see www.generative-modeling.org). Its main purpose is to serve as a general exchange format for procedural models, e.g. as a file format for encoding the construction history of complex objects. It is capable of generating large amounts of geometric data out of very compact descriptions. Since the syntax is very similar to Adobe's PostScript language, GenerativeML can be thought of as a kind of "3D PostScript".

The pipeline is focusing on common features found in the subsurface infrastructure. Utility infrastructure, like the underground water or gas distribution systems are arranged in traces, divided in multiple layers at different depths. Depth is either given as an attribute value, or must be estimated based on heuristics (e. g. telecommunications are typically in the first layer at 0.5m depth). The depth at which each utility system is buried also depends on the challenges of the terrain.

The central object considered in the transcoding pipeline is the pipe. All types of pipe-shaped infrastructure, be they electricity, gas, water, sewer or heating pipes are abstracted by the same common geometric attribute. However they can be visually discriminated given their semantic attributes which can be exploited with style maps. One pipe can consist of several segments. To avoid the creation of excess polygons, the transcoding deletes collinear points as well as duplicate points automatically. It is also possible to quantize coordinates for LoD generation since millimeter level precision is typically not required in the visualization.

The non-geometric attributes such as id, purpose or ownership are converted to semantic mark-up, while the geometric attributes and the radius are passed to the GenerativeML. Traces and layers are described analogously to pipes but with a rectangular rather than a circular cross-section. In addition to pipes, the underground infrastructure consists of special facilities, e.g. gate valves, water hydrants, T-fittings, and so on. These facilities are blended with the generated 3D environment using a 3D model library based on special rules. Buildings are included to provide geographic context. Their footprints are transcoded by simple extrusion, using either a height attribute or a default value if height is unavailable. For more detail of the used GeographyML format please refer to the specification of VidenteGML (Junghanns, Ranzinger, Schall, & Reitmayr, 2010).

The transcoding pipeline is implemented in C and follows a very simple structure, as mainly string operations are performed to convert from one format into another. The transcoding pipeline has the responsibility to generate one or more Open Inventor files used for visualization. The Open Inventor file consists of one node for each feature containing all properties and geometry descriptions of the feature. Depending on the feature type, the according scene-graph node is generated. The properties of a feature are transferred into a key-value pair. The geometry is written into a subnode.

For features of type *pipe* the nennweite-property value is extracted and used for calculating the diameter or radius of the pipe. For features of type *extrusion* the height is set to a predefined value. Features of type *model* are provided with the appropriate link to a 3D library storing the 3D content. For *trace* features the breadth-property value and the depth-property value are used to calculate the dimensions of the feature. From *terrain* features a 3-dimensional grid representing the DTM is generated.

Once all features in a GeographyML file are transcoded into Open Inventor nodes the data can be used for 3D visualization. An example node for a feature in Open Inventor format is shown in Appendix 9.3 in Figure 87 and in Figure 88 including GenerativeML parameter descriptions.

The configuration file supporting the transcoding process stores a specific set of rules to handle the following issues:

Relationship. Define the relation between features described in GeographyML and nodes in Open Inventor.

Filtering. This is responsible for deciding which feature geometries and properties from the GeographyML side are used.

Grouping. Single features (e.g. gas, electricity or water) that have not been grouped on the GeographyML side are grouped by the transcoding pipeline. At first, features are grouped by category like gas, electricity etc.

Legacy style. Styling information is added on how single features should be visualized (e.g. gas pipes are visualized in green color).

Data cleaning. This is necessary since a lot of data is generated in the transcoding pass. Optimization strategies like LoD reduction or polygon count reduction can be applied to simplify the resulting data. Moreover, consecutive point that appear twice, so called double points, are removed.

Tracking. Necessary information used for tracking is added (e.g. information for geo-referencing the data). Most of the data provided usually comes on decimal precision at a millimeter level. Furthermore, it may come on different metric systems. This descriptive information is attached as contextual attributes inside the GeographyML file, and may be used for data cleaning. This poses the simple question what contextual information should be forwarded to the scene-graph and which one should be filtered by the transcoding pipeline. To solve this, a configuration file out of the input GeographyML file is generated. This file allows the user to effectively decide which data to forward, ignore, or use as keywords for filtering during the transcoding step. Additionally, the separation of styling management allows the creation of autonomous styling tools independent of the data to be visualized. This enables an effective platform for visualization of heterogeneous data.

4.4.2 Anatomy of the geospatial infrastructure

The pipeline is able to process the most common features appearing in the underground infrastructure. In the following these features are outlined.

Trace. Individual utility layers, like the underground water or gas distribution systems are arranged in traces.

Layer. A trace can be divided into one or more separate layers. Each layer is situated at a different depth and has the same width as the trace. The depth at which each utility system is buried depends on the challenges of the terrain. Usually the first layer below the surface consists of telecommunications and electrical cables. Down another layer at about 1 m depth gas pipes are buried. At 1.5 m depth there are water pipes followed by a layer 3 m underground consisting of sewer mains. At the moment no DTM has been incorporated, but will be left as future work.

Pipe. The transcoding pipeline is only dependent on a GeographyML application schema. Any type of information encoded in a GeographyML format may be transcoded regardless of their semantic interpretation. Any type of information encoded in geometric attributes may be translated, for example, electricity, gas, water, sewage and heating pipes are all abstracted by the same common geometric attribute. However they can be visually discriminated given their semantic attributes to which styling techniques are applied.

Figure 25 shows the two possible output visualizations of the transcoding pipeline. A simple Open Inventor scene-graph visualization that simply draws single line segments is compared to a GenerativeML visualization. A major contribution of the work is the simplicity of the interface between the scene-graph and the model generation using GenerativeML. It can be seen that one pipe can consist of several segments. Small rings around the pipes represent pipe collars connecting pipe segments. These ring elements are only visual aids to instruct the use that an "elbow" might be present in that location. These are always located whenever there is an angle in the pipe coordinates. If a pipe had only collinear coordinates, these would not be decorated with rings. For each pipe both geometry and property attributes exist as input to the pipeline. The geometry of a pipe is described by the 2D coordinates and a depth attribute. Additionally all attributes of each pipe are delivered by property attributes. To avoid unnecessary polygons, there is the option to delete collinear points as well as double points automatically from the geometric coordinates. The parameters for creating the 3-dimensional model of the pipe are calculated as follows:

The coordinates of the junctures are reduced in precision and then forwarded as a coordinates list to the GenerativeML nodes (Havemann & Fellner, 2004). Their precision is reduced from millimeters to meters to avoid aliasing artifacts in the final image.

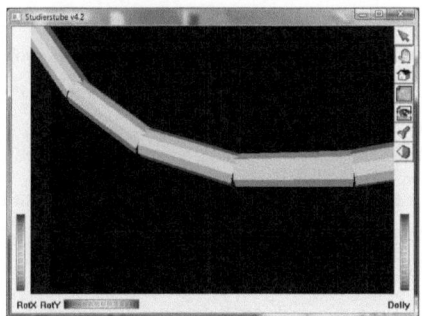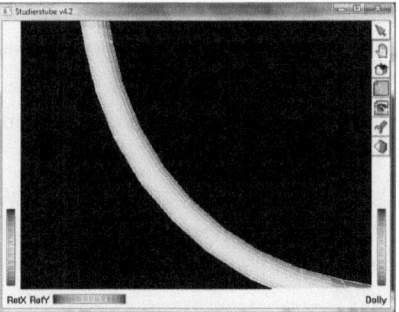

Figure 25: Transcoding output formats. (Left) Curved pipe in Open Inventor format. (Right) Curved pipe in GenerativeML format.

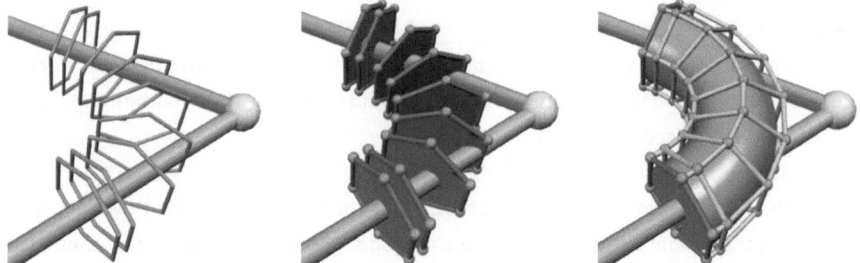

Figure 26: Curve geometry. To achieve nice rounded edges, a circle segment is computed to connect subsequent tubes. The circle segment is offset to either side and sampled to obtain the radial line segments (left). The circular profiles (n-gons) are converted to double-sided faces (middle) that are then connected using a make-tunnel operation (right).

Pipes are two dimensional entities, and therefore all its junctures are at the same height. This height is forwarded as an extra single attribute, which then gets used by the GenerativeML interpreter. The radius of pipes is also extracted from the semantic attributes and forwarded along the geometrical attributes. Also the thickness is provided for every pipe, but this is fictional since no record of the real thickness is provided. In this case 10% of the pipe radius is used. The ring added to signify a pipe juncture has a thickness of 20% of the pipe radius and its length is 10% of the shortest segment in this particular pipe. Every pipe (and also trench) can have an opening angle, but this is set during runtime given the semantic attributes of the object. An example image of this is provided separately (see Figure 26).

The great power of GenerativeML is that it supports process chains quite efficiently. The stack is a flexible way for passing data produced by one function as input parameters to the next. This is exploited to create different types of pipes simply by passing different profiles to a connecting function (see Figure 26). To open up the rounded pipes (openangle parameter), two sampled circle segments with different radii are connected. The same way pipes with rectangular profile can be created, so called trench style (2D parameter trasserad for x/y-dimensions), and these trenches can also be opened up.

Trench. Trenches are described analogously to pipes but with a few differences. Instead of a radius property a trench has width and height properties. All the attributes such as ring thickness that depended on the pipe radius depend instead on the trench height.

Model. Additionally to pipes, the underground infrastructure consists of special features, which, for example, are used to interconnect the pipes or ducts to access the pipes. There are features such as gate valves, slide valves, water hydrants, T-fittings, and so forth. In the GIS they are usually stored as symbols (see Figure 28). A 3D model is stored in a library per symbol. The geometry property of the feature allows extracting the position of the feature in the underground infrastructure. The orientation property of each feature is used to orient the 3D model correctly.

Extrusion. Typically, in GIS buildings objects and the like are described by their footprints. The transcoding pipeline allows to transcode such features too. In this context all attribute and geometry properties are written to the output file. The footprints can then be extruded. Unfortunately, most of the information on buildings does not include height information. Therefore, at the moment only synthetic values are used for the heights of the buildings.

This poses the simple question what contextual information should be forwarded to the scene-graph and which one should be filtered by the transcoding pipeline.

To solve this, a configuration file is generated out of the input GeographyML file. This file allows the user to effectively decide which data to forward or use as keywords for filtering during the transcoding process. The result is a scene-graph description with per-feature grouping of shape objects and semantic attributes. This has the advantage that in order to highlight all objects of a certain type inside a scene-graph (e.g. by changing their color to red) the respective material properties of all affected nodes must be changed somehow. The strategy is to change only one "style node" for all the desired nodes early in the traversal order. Each affected node updates its styling because its attributes have been touched (Mendez et al., 2008). This allows for changing the appearance of the geospatial model of the environment on the fly.

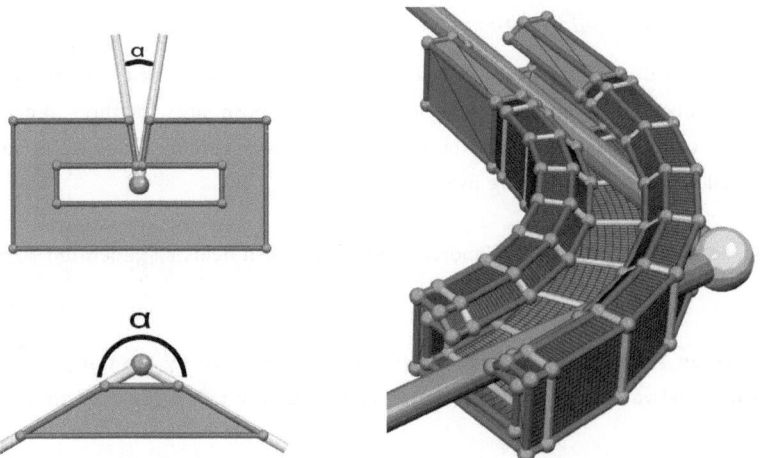

Figure 27: Trench with different opening angles. Trenches or pipes can be modeled with an opening angle to be able to look inside.

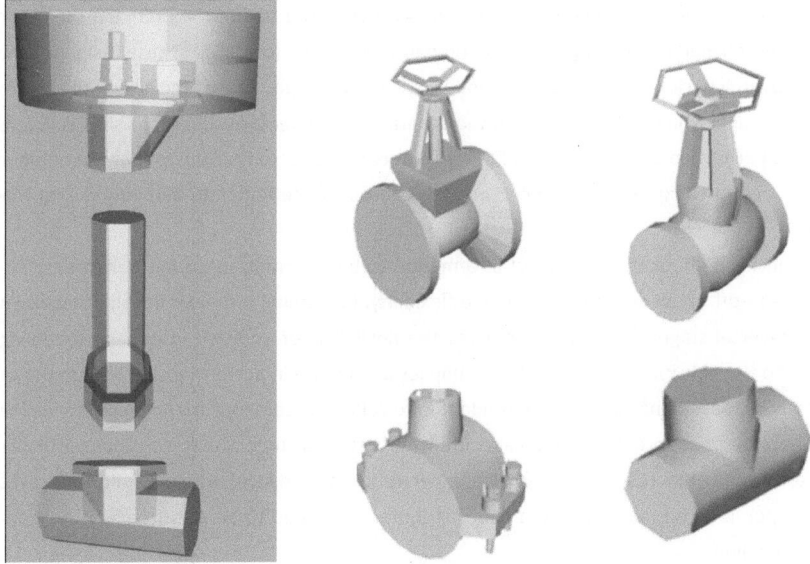

Figure 28: 3D models representing GIS symbols. Examples from the self-made library of objects such as water hydrant, water shut off valve and socket.

4.4.3 Transcoding trade-off analysis

In order to study the effects of the *transcoding trade-off*, a series of tests have been performed involving different pipe networks as well as mesh sizes for three separate stages of the transcoding. To assess the performance the according times were measured. The following conditions were used in the tests:

S0: Static. Semantic attributes are ignored; one single mesh holds all generated 3D objects.

S1: Adaptive on Transcoding. Semantic attributes are evaluated by the transcoding to generate material values for every geospatial feature before being deployed to the 3D browser.

S2: Adaptive on Traversal. Semantic attributes are preserved and evaluated during scene graph traversal where template mapping and material bindings take place.

The S1 condition performed on average at 64% compared to S0 (σ = 1.05). On the other hand, S2 performed on average at 57.5% compared also to S0 (σ = 1.55). These tests indicate that the overall performance level of S1 and S2 will remain regardless of the size of the mesh; the performance will mainly depend on the separation of objects in subgraphs. As expected, S1 and S2 performed at similar rates (6.5% difference). The reason for this is that the number of traversals is the same; the overhead on S2 is caused by the template mapping during traversal.

The decision about the amount of information to be reformatted during transcoding implies a trade-off between performance and flexibility. Preserving semantic information down to the traversal stage significantly increases the flexibility for applying visual and modeling changes to the objects in the scene. It also implies a decrease in performance. The number of traversals increases since every node is adapted to reflect its semantic mapping by traversing a styling node prior to the geometrical content. A better strategy was to consider particular user tasks (as in the case of infrastructure network maintenance) which would reduce as much as possible the number of semantic attributes that need to be preserved by the transcoding pipeline.

Transcoding experiments were performed with urban areas of the size up to a few square kilometers. That is the typical size of a dataset needed for industrial tasks. Next, selected transcoding results on a UMPC (Sony Vaio UX, Intel Core Solo 1.1GHz) are presented. Figure

29 shows a 3D model of an urban area of Salzburg including extruded building footprints and underground infrastructure. The overall number of features in the dataset is 357 including 348 pipes with 2659 pipe sections and 9 extrusions. The transcoding of this dataset on a UMPC (Sony Vaio UX, Intel Core Solo 1.1GHz) took 0,772 seconds. Figure 30 depicts a 3D model of an urban area including extruded building footprints and underground infrastructure. The overall number of features in the dataset is 398, including 378 pipes with 2322 pipe sections and 20 extrusions. The transcoding of this dataset took 0,781 seconds.

Furthermore, Figure 31 shows a visualization of an urban area in Graz with 682 features. The model includes the following features: extruded building footprints, fences, parcel borders, gas pipes, water lines, low power electricity lines and capping street features. The according transcoding time is a bit more than one second. Generally, transcoding time increases linearly with the complexity of the dataset, respectively the number of features and the number of sections of features. Table 1 lists the according performance numbers.

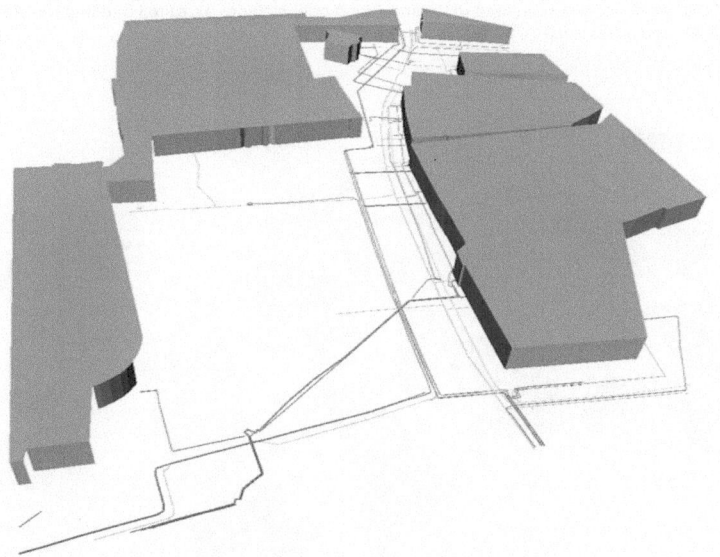

Figure 29: 3D model of an urban area in Salzburg, i.e. the area around the Residenz. The model includes extruded building footprints and underground infrastructure networks.

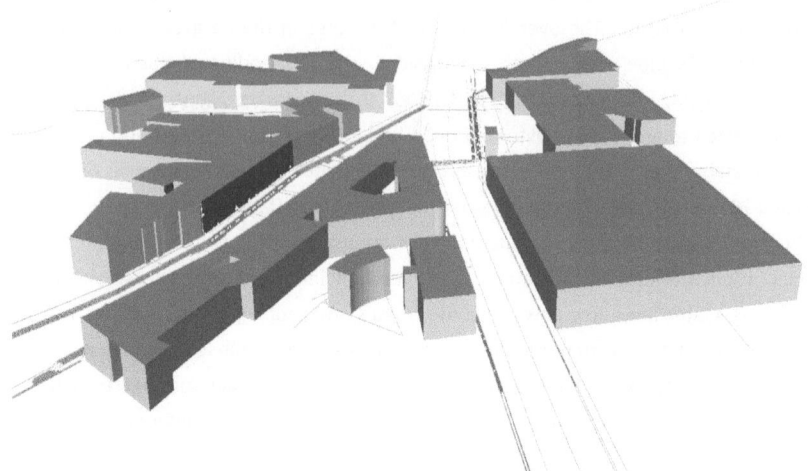

Figure 30: 3D model of an urban area in Vienna. The model includes extruded building footprints and underground infrastructure networks.

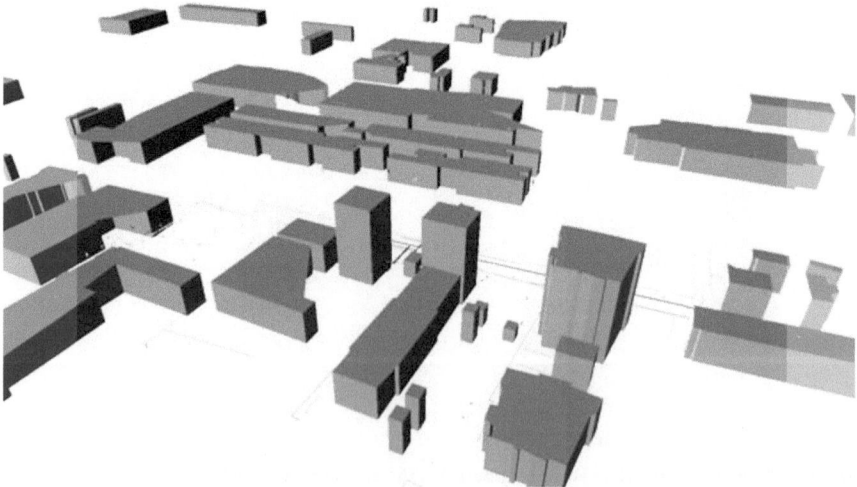

Figure 31: 3D model of an urban area in Graz in Sandgasse/Inffeldgasse. The model includes extruded building footprints, underground infrastructure networks and street-level features.

Table 1: Transcoding results of three different urban areas.

	Geospatial model 1	Geospatial model2	Geospatial model3
location	Salzburg Residenz	Wien Marianneng.	Graz Sandgasse
# of features	357	398	1163
# of pipes	348	378	682
# of pipe segments	2659	2322	5388
# of models	0	0	404
# of buildings	9	20	77
transcoding time [sec]	0.672	0.781	1.178
# double points removed	459	2	0
size of area [m2]	350x350	500x500	1000x1000

4.4.4 Limitations

The transcoding pipeline has a few limitations. Its information flow is strictly one-way, from the scene-graph to the GenerativeML nodes that generate the geometry procedurally. The scene-graph is static as at runtime, no nodes are added or deleted; only the connections between these nodes can be changed. More flexibility could be obtained by creating parts of the scene graph procedurally.

For large networks, it is desirable that the scene-graph can be loaded progressively: nodes are refined by inserting a subgraph or subgraphs are collapsed into a single node based on proximity and visibility as well as on semantic queries.

4.5   Discussion

Three approaches for generating geospatial modeling have been shown. In contrast to the manual and semi-manual approach, a method based on transcoding legacy data seems most promising for generating semantic AR models. Most importantly, the transcoding approach allows for the efficient generation of large-scale models. The simple navigation example at the beginning of the section illustrated that semantic information of the model is used for assisting the user in the task, e.g. re-calculating the path through a building. Moreover, the example demonstrated that the geometric information of the model is not only used for visualization purposes but also for tracking fiducial markers, as the application derives the tracking data from the BAUML model. This helps to understand the requirements for AR models.

Taking these requirements into account, the transcoding pipeline contributes to an automatic generation of 3D models for AR from existing legacy data sources. The transcoding approach demonstrated that models for AR can be created automatically and efficiently. One on the main strength of this approach is the possibility of using procedural models. An advantage is that complex objects can be represented with a small number of parameters. The description is very compact and would allow very short download times via a wireless link. Furthermore, the visualization of the procedural models can be very detailed and impressive. But, this also affects the rendering performance. Consequently, depending on the application's needs, specific objects will be represented as a procedural model and others will be represented as simple Open Inventor models.

A display connected to a digital computer
gives us a chance to gain familiarity with
concepts not realizable in the physical
world. It is a looking glass into a
mathematical wonderland.

Ivan Sutherland
Computer graphics pioneer,
vice president Sun Microsystems

5.   Hardware setups for augmented reality

The author had the chance to experiment with a large variety of sensors based on mag-netic, inertial, infrared and UWB (ultra-wideband) physical principles. With such sen-sors, the author developed hardware platforms to satisfy the requirements for the spe-cific AR applications. This chapter deals with mobile and handheld setups the author has (co)-developed. An essential aspect of AR is the ergonomics of the device and its user interface, especially for outdoor environments. Various hardware setups that were built in order to gain experience with different tracking systems and applications are briefly overviewed here.

The successful delivery of mobile AR is an ongoing challenge as interactive 3D applications must be implemented on limited hardware platforms, requiring tracking over a large area of operation at high accuracy. When looking at the evolution of mobile AR setups, one can ob-serve that over the last decade AR was feasible on continuously smaller devices such as UMPCs and modern smart phones described by (Wagner & Schmalstieg, 2003). Today, smart phones represent low-end setups since the built-in sensors and processing power mirror the capabilities of AR setups of a decade ago. In contrast, the aim of the setups presented next was to build high-end configurations using state of the art sensors which allow for the most advanced tracking quality. Early experiments with backpack setups using HMDs are not de-scribed. The focus lies rather on see-through displays based on mobile and handheld devices.

The author's interest was in experimenting with smaller handheld computer platforms which allow a "magic lens" style of video see-through augmentation. Such a handheld AR platform is inexpensive and ergonomically superior to the backpack solution. Most potential users are already familiar with camcorders and consequently understand the use (hand-eye coordination) of a handheld video-see through device. Subjectively, the author has observed that users prefer handheld AR over head mounted displays despite the lack of stereoscopic graphics and hands-free operation. The lower computational power of handhelds is partially compensated for by the reduction in graphical complexity: monoscopic rather than stereoscopic, smaller screens, increased tolerance for lower resolutions.

5.1   Requirements

Mobile AR platforms should provide a maximum of performance and usability while minimizing weight and size. Outdoor AR applications are particularly challenging in terms of hardware requirements. There is no room for placing permanent instrumentation in the environment, thus the AR platform needs to be completely self-contained and fulfill aspects such as sunlight-readable display. A system of any practical value must at least address the following challenges:

– The system must provide sufficient computing capabilities on a platform that allows for several hours of battery-powered operation.

– The system must have an ergonomic form factor that allows holding the AR device for extended periods without excessive fatigue, and performing typical operations with high convenience.

– The system must include a pose tracker which delivers six-degrees-of-freedom with real-time updates, is globally registered and robust.

Compared to body worn equipment, a handheld device is less intimidating and can be more easily shared by multiple workers. A handheld AR display — as opposed to a head-mounted display — can also be viewed collaboratively. Setups for two types of computing hardware, namely UMPCs and tablet PCs, were investigated.

5.2   UMPC-based setups

The UMPC is extremely powerful given its weight, but with the additional peripherals required by AR, the weight adds ergonomic restrictions on the duration and type of actions being performed. Hence, new devices were constructed that are described in the next section.

In 2006 the author has built an experimental prototype handheld system, consisting of a Sony VAIO U70 and a variety of different sensors attached to an acrylic mount Figure 31 (left). The sensors included an Infrared tracker as well as a magnetic FOB tracker, which were typically used for VR applications. Figure 33 shows a further developed version of the setup. The sensors consist of a USB camera serving the dual purpose of providing images for an optical tracking system and also for providing the video required by the "magic lens" metaphor; a Ubitag, providing position estimates only; and an Intersense InertiaCube3 inertial tracker providing orientation estimates only. All sensors were connected via USB hub with the USB port of the UMPC.

An AR setup for outdoor use can simply be build by adapting the indoor UMPC setup, integrate new sensors and write the appropriate software drivers.

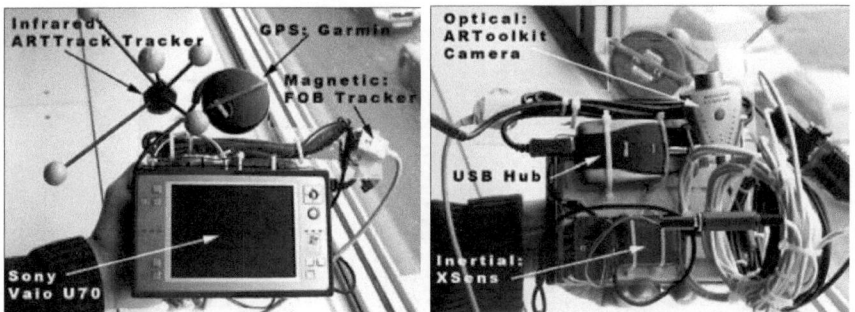

Figure 32: Unconventional indoor AR setup. Sensors are simply fixed on a plexiglas plate using wire traps (front and back view).

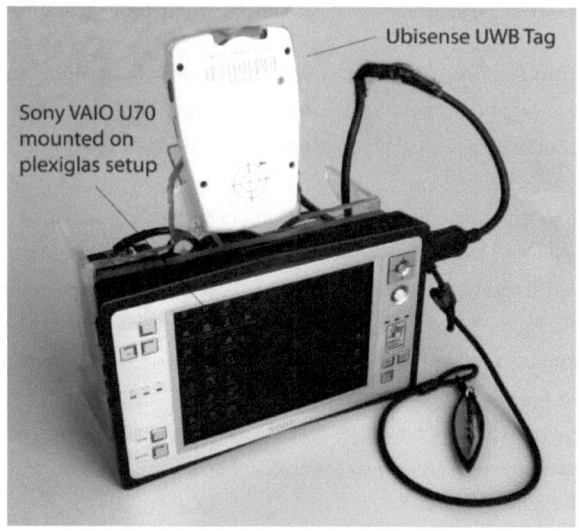

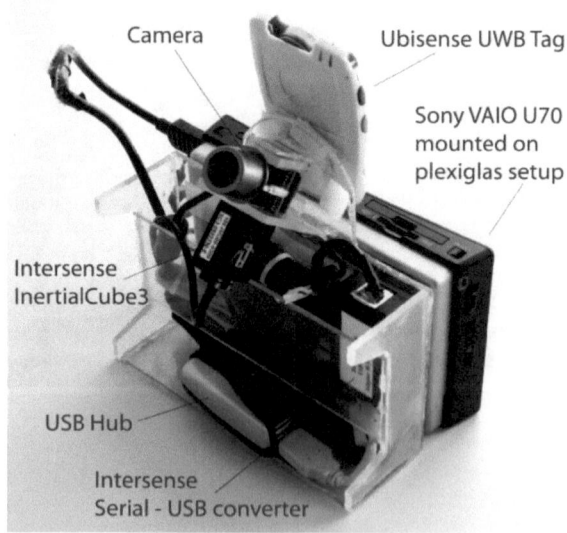

Figure 33: Indoor AR setup. (Top) Front view of the mobile device.
(Bottom) Back view of the mobile device.

5.3   Vesp´R setup

After building these UMPC-based setups the idea arose to design a more experimental
setup allowing for better ergonomics, form factors and ways of interaction. With this in
mind and considering the requirements for mobile AR setups, (Kruijff & Veas, 2007) designed a two-handed shell around an UMPC (Sony Vaio UX, Intel Core Solo 1.1GHz,
Windows XP, 0.5kg). The shell was manufactured from lightweight ABS plastic, and holds
the sensors. Moreover, the system used the *Studierstube* software framework
(Schmalstieg et al., 2002).

In order to come up with a suitable device construction, an extensive design study was
performed finding ways to support mobile spatial interaction using a handheld in new and
effective ways. As a result of this analysis, the following needs on the construction were identified: it needs to hold the additional peripherals, make available a range of well reachable
controllers, and allow for flexibility in usage including freedom of movement. The latter specifically requires that multiple grips are possible: one possibility in which the construction can
be grabbed with both hands to split the weight on both hands and arms, and a single-grip
version that allows for the second hand to either control the UMPC (pen input) or perform an
independent task (like marking a road).

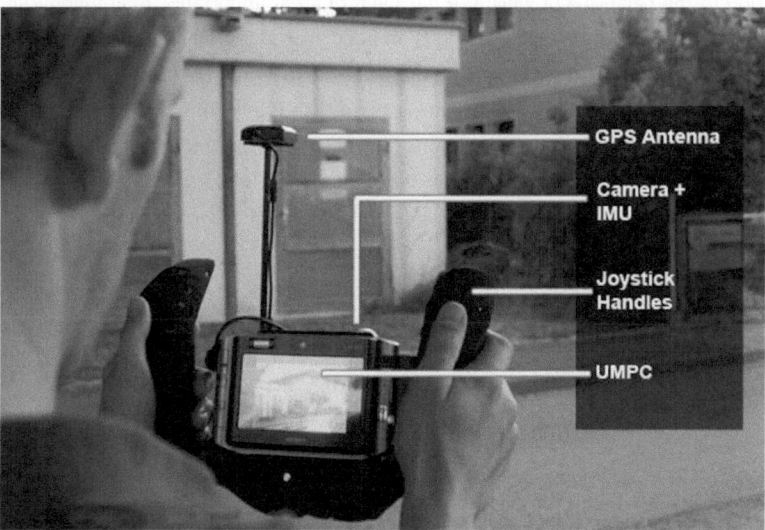

Figure 34: Outdoor AR setup named Vesp´R. It has been designed for an ergonomic handheld
AR device around a UMPC.

After the study, multiple prototypes were created to come to the final construction called Vesp'R. The Vesp'R construction, made of sturdy ABS material covered with rubber, consists of a main hull around the UMPC, to which either one or two handles can be connected (see Figure 34). The hull contains an empty space, holding the peripherals (GPS, orientation sensor and camera). The handles, also simply called "joysticks" hold multiple kinds of controllers, from simple micro-joysticks to midi-components. Currently, the application mostly makes use of micro-joysticks and buttons.

The first setup consists of two handles connected to the sides of the hull. In this way, weight is equally distributed over both hands, and can be handled well due to the powerful grip on the joysticks. Hence, users can make use of the device construction for longer periods of time without being restricted by fatigue that is possibly caused by holding the construction in front of the body.

In the second configuration, the joysticks are removed from the side: one joystick is placed below the hull. Due to the power grip on this joystick, which supports a steady way to hold the construction, it is possible to make use of the second hand for other tasks. However the single-handed grip can cause fatigue in the arms and hands: this setup cannot be used continuously for longer durations. However since the second hand can always support the construction, periods of relief can be added to the task performance cycle.

In addition to the balancing of weight, it has been of great importance to ergonomically map functions to controllers. The power grip is an important factor while using Vesp'R: the fingers grasp the handle and press it against the palm of the hand. Hereby the thumb and index finger can be moved freely when balancing the device firmly. This ability is the key to the control structure because all major tasks are controlled with the thumb and index finger. For this purpose an eight-directional micro joystick is mounted in one of the handles easily accessible by the thumb. On the back of the grip a second joystick with a trigger button is placed, which can be used with the index finger.

Menu control with the micro joystick is kept straightforward. For mode changes, a linear menu overlay is operated with the left/right direction of the joystick. Similarly, the micro joystick is used to perform constrained spatial interaction, such as moving objects in the ground plane.

5.4   Evaluation of Vesp'R setup

In order to get an overview of the quality of the different system components, a structured evaluation, analyzing a range of aspects was conducted. These factors included the general quality of the user interface, the visualization methods being used, the matching of industrial requirements obtained in the system requirement phase, and the actual operation by end-users including cooperation aspects. The factors were mapped to different kinds of users in order to get the most useful feedback. First, system aspects were examined with system experts in order to go through refinement cycles of the system before the system was presented to the actual end-users. Naturally, end-users had seen and used prototypes of the system before: their feedback has flown directly in the initial development phases.

5.4.1 Mobile computing developers

The first range of trials and interviews was performed at the Ubicomp 2007 conference. Ubicomp is the one of the main conferences visited by mobile computing experts. The demo setup focused on analyzing an underground infrastructure of the area around the conference centre. The main aim was to investigate the quality of the software and hardware interfaces to control the application. In the evaluation, users made use of the single-handed setup of Vesp´R. Following the usage sessions, participants were requested to fill in a questionnaire using a 7-point Likert scale rating.

17 participants (16m / 1f) took part in the experiment. All had good computer skills, but no professional experience with GIS for underground infrastructure. The average duration of the session was about 10 minutes per person. At the beginning of the session, each participant was instructed how to interact with the application using Vesp'R. Participants could see the underground infrastructure superimposed in 3D on the street below a platform outside of the conference centre (see Figure 35). During the try-out, complete freedom was given to the participant to orient the device and look at different parts of the underground infrastructure.

The participant was also asked to use the controls of the Vesp´R device for interaction like switching on/off single layers of underground infrastructure. Furthermore, the user could switch between different visualization styles like X-ray or excavation and also combine both to avoid display clutter and use the Vesp´R for further interaction with the geospatial data. After finishing the session, the participant rated specific aspects concerning both the device and the user interface. The questions were grouped around two main topics: hardware setup (including ergonomics) and user interface quality.

Overall, the participants were relatively satisfied with the placement of the controllers, and did not perceive fatigue.

Figure 35. A participant testing the Vidente application exhibited during the Ubicomp 2007 conference.

The weight balance of the device was not rated satisfactory for most participants. However, most users did not report fatigue. Unfortunately, the setting did not permit comparison against other grip configurations, holding a standalone UMPC, or even holding a full-size laptop such as currently used in conventional field work.

Observations showed that a significant portion of the users held the device single handed. Those who used two hands placed the non-dominant hand in the round back of the device without having received explicit instruction how to hold it (see Figure 35). One of the users held the device always from the back and only accessed the control buttons sparsely with his dominant hand. None of the users showed any significant hand tremor or high muscular tension; only one of the users had to lay the device down for a couple of seconds to relief strain.

During the try-out, direct sun cast led to rather poor display contrast causing most users to hold the device at eye level instead of chest level as expected. It is likely that the uncomfortable pose affected the subjective rating of ergonomics. Subjects rarely switched the focus of their gaze from the on-screen image to the real world, suggesting that the depth cues of the application were sufficient to provide spatial awareness. The effectiveness of the application control was also received positively, while the usefulness and effectiveness of the controls received high acceptance.

5.4.2 Mixed user group

Obviously, there were mixed results from the first evaluation, in which most people did not find the single-handed grip satisfactory. However since there was no comparison to other kinds of grips and device setups, the authors did not know the actual value of the results.

Hence, in order to get a better insight on the ergonomic factors of the Vesp'R setup, a study was performed to focus on the pros and cons of the different possible grips in comparison to other device setups. In the test, subjects had to perform a placement task in which objects needed to be moved virtually from one location to another, thereby applying different kinds of grips and (body) postures to use the devices. The following device setups were used: UMPC only, UMPC with simple plastic enclosure, single-handed Vesp'R mode and two-handed Vesp'R mode.

15 subjects (12m/3f) participated in the evaluation, having different "body configurations": people were picked with different hand sizes and levels of "muscular conditions" (normal people, sporty people) to see how easy it was to hold and control the different setups. All users had a background in computer science, but this had no direct effect on the ergonomic considerations being evaluated. Figure 36 depicts results of user comfort ratings of two-handed usage and single handed operation. The different setups have largely varying weight factors: the UMPC only just weighs about 550 grams, whereas the Vesp'R with two joysticks and all the needed peripherals gets close to 1250 grams. As such, it came to no surprise that most users found the UMPC comfortable to hold (avg. 5.33/stdev 1.54) – nonetheless, the two-handed Vesp'R was rated most comfortable, which was a big success considering the weight difference with the other devices (avg. 6.07/stdev 1.38). Users could easily balance the weight (avg. 6.20 / stdev 1.20) and found the controllers well-placed (avg. 6.00/stdev 0.88). In line with the first experiment, the single-handed version of Vesp'R was rated less comfortable, but still within mid range (avg. 4.60/stdev 1.80): users could still hold the construction.

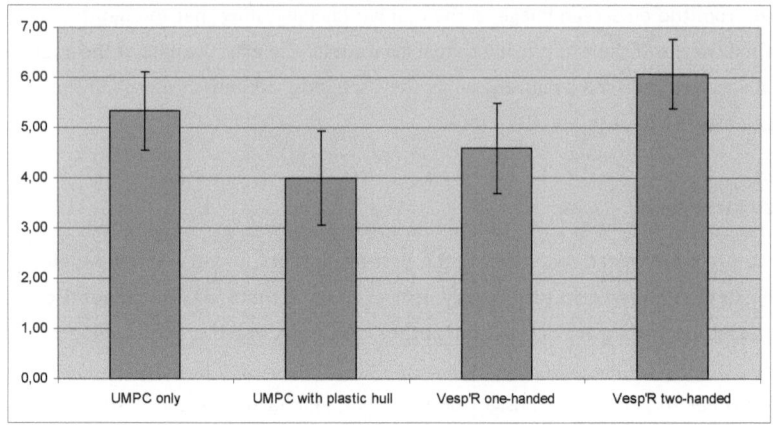

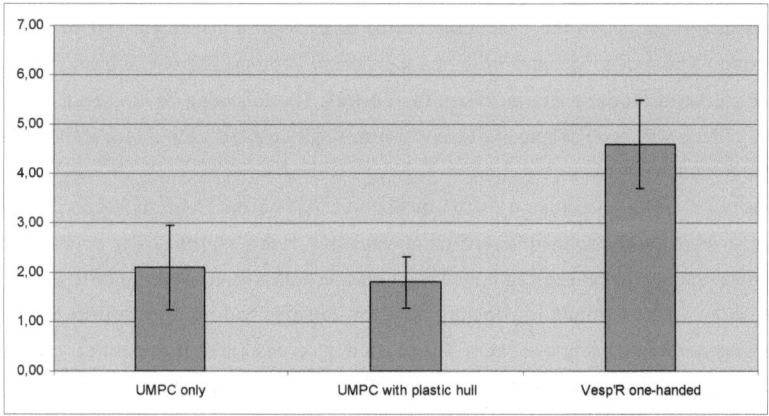

Figure 36: Evaluation results. (Top) User comfort with two-handed usage. (Bottom) User comfort with single handed operation.

It is not the user's first choice, but, in case the second hand needs to be used for another task it easily outperforms the UMPC setup: all users did not prefer to make use of the UMPC in single handed mode, since it easily tilts to one side, causing considerable fatigue.

In relation to user comfort and the placement of the controllers, users found the two-handed Vesp'R the best choice for interacting with an application (avg 6.20 / avg 0.56). Both the UMPC and the UMPC with plastic hull only scored mediocre in the range around avg. 3.60. Users also found they could still operate an application single-handed with the Vesp'R (avg. 4.60 / stdev 1.64), whereas they could not at all perform interaction with the UMPC and

UMPC with plastic hull configurations (both rated around avg. 2.00). These results were completely in line with the user comfort results.

Overall, the study showed that the two-handed Vesp'R is ergonomically superior to all other setups and can be well used for longer durations. The single-handed Vesp'R is not the ideal choice, but currently offers the only acceptable solution for mixed tasks: UMPC only / UMPC with plastic hull configurations are not suitable.

5.4.3 Field worker interview

The previously presented studies mainly covered ergonomic issues of the hardware setup, whereas the following studies focus on the practical relevance of the prototype. By conducting an interview with field workers from local industrial utility companies (two employees from the local power supplier E-Werk Gösting Stromversorgungs GmbH and 3 employees from ENERGIE GRAZ GmbH & Co KG) valuable feedback from experts with strong practical experience for years was gained. Four employees had significant background in the electricity sector whereas one employee had relevant experience in the gas supply sector. First, the Vesp´R setup (single handle as well a two handle setup) and the Vidente application were introduced to the field workers. Then they used the setups to visualize the underground infrastructure at the outdoor site and interacted with the application (see Figure 37).

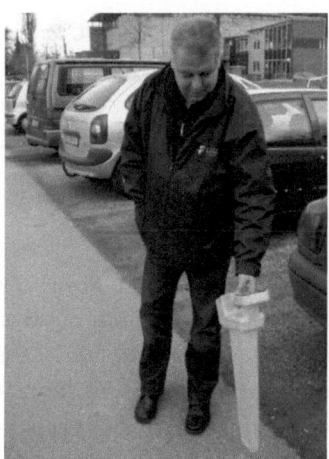

Figure 37: Expert field worker. (Left) Expert is using conventional measurement device and (above) two handle Vesp´R setup for finding underground infrastructure.

Second, a semi-structured interview was conducted to assess the practical applicability of the setup and the application.

Scope of application. The interviews showed that field workers from both companies E-Werk Gösting and ENERGIE GRAZ gave positive feedback to the prototype. They confirmed that potential fields of application are tasks like construction instruction, outage management and planning.

Hardware setup. A basic question was if field workers prefer HMDs to handheld devices. It turned out that handheld AR devices are rated better, because they provide a "magic lens" rather than high immersion. Because of optimal characteristics in terms of balance and weight, field staff from E-Werk Gösting preferred the Vesp´R hardware setup with two handles clearly to the one with the single handle. Generally they were satisfied with the interaction capabilities of the setup. But two field workers expressed the wish for further interaction possibilities, like a scroll wheel. Slightly different, field experts from ENERGIE GRAZ considered Vesp´R with two handles and Vesp´R with one handle equally useful. They remarked that the latter setup allows for spraying markers on the ground at the same time. A major issue for outdoor use concerns the ruggedness of the system, which is water-repellent casing, sunlight-readable display, well protected sensors and grip material suitable for rough outdoor conditions.

Operation mode. Lately field workers at ENERGIE GRAZ started using a tablet PC for the process of marking gas pipes in order to test for leaks in a yearly interval. Therefore GIS data is stored on the mobile tablet PC and is synchronized with the office GIS system weekly. E-Werk Gösting usually needs to locate 50-100 meters of trench length a day. The device would be operated in a discontinuously mode, which means the field worker uses the Vesp´R setup for approx. 5-10 seconds, walks further, and again uses the device. The overall time of usage at one construction site would be around 15-20 minutes.

Field workers form E-Werk Gösting mentioned the system would alleviate their work by allowing them to carry fewer measurement devices with them. Commonly a smaller workload and less wrong excavations are expected by using Vidente. The biggest advantage of the system is an improved spatial overview of the construction site through the egocentric visualization of the underground infrastructure. All field workers clearly preferred a 3D visualization to one in 2D. In particular, the depth perception of the pipes was considered beneficial. Vesp´R could also function as a device for measuring the position of a newly passed pipe by simply following the path of the pipe and stopping at several positions to record the accord-ing GPS position. In this way — given high tracking accuracy — the position of new pipes could be measured accurately.

5.4.4 Management level feedback

In the final stage, feedback was gathered from industry at the Austrian Smallworld User Group Meeting 2007 (ÖSWUG) where attendees (management level, field workers) from approx. 20 utility companies were present. The hands-on outdoor demo (using the sin-gle-handed Vesp´R setup) visualizing the underground infrastructure at a nearby street crossing was shown. Seven people answered a short questionnaire after the outdoor test. Since most people were at a management level, useful feedback from a manage-ment view from utility providers to this application was obtained.

Only some rated the weight and form factor of Vesp´R no as not optimal, which is in line with the previous results on the one versus two-handed setups. When asked for the advan-tages they mentioned the visualization itself, the interaction with the application and the real-time tracking. Attendees rated the usefulness of the visualization as high. Significant time savings could be achieved using a system like presented. Many providers could foresee using this application in the process of construction instruction. Furthermore, people expressed the wish for seeing 3D city models additionally to underground infrastructure.

5.4.5 Evaluation summary

The evaluations provided a wealth of information. Overall, the interactive visualization seems to be appropriate for the end-users: both field workers and management claim that the presented methods provide an effective and highly useful method for outdoor inspection tasks, probably saving both time and money. Hence, the interactive visualiza-tion is in line with their industrial and operation requirements. Though sometimes re-ceiving mixed results of computer scientists, end users were positive on the developed hardware infrastructure, not too much worried about fatigue effects as the authors were afraid of. Better said: the device construction in both one and two handed ver-sions could match the ergonomic requirements for most tasks of the field workers. The one handed construction may not be ergonomically ideal, but a big step forward in comparison to older setups and allows for user freedom in performing non-computing tasks. Notwithstanding, the system still has potential for improvement, both on the hardware and software side, to even better support the user needs.

5.5   POMAR-3D setup

Having done evaluations and interviews about the Vesp´R setup, the author noticed that Vesp´R is a nice approach for experimenting with mobile spatial interaction. But, in terms of robustness and outdoor roughness the design was not satisfying. Moreover, field workers rated the Vesp´R setup as cool and interesting for mobile spatial interaction experiments. But field workers currently employ tablet PC-based setups. Consequently, a tablet PC-based AR setup would be more realistic for industrial use. For a project named POMAR3D that aimed at high-precision tracking for outdoor AR, in a next step a handheld AR platform was developed that is build around a Sony VAIO UX UMPC with 1.06 GHz Pentium CPU based on the work of Kruijff et al. (see Figure 34). The setup was designed for more robustness and better shielding the sensors from harsh outdoor influences. Alternatively, the author employs the rugged UMPC Panasonic U1 featuring sunlight-viewable touch screen and sealed all-weather design for demanding outdoor use.

The handheld platform is equipped with various sensors. A UEye USB 2.0 camera using a 4.2mm wide-angle lens provides the video-background and delivers the video frames, which are used as input for the visual tracker. The camera captures video frames with a resolution of 640x480 at 30 Hz. A 3DoF inertial sensor (XSens MTi-G, built-in GPS receiver is not used) is mounted at the encasing at the back of the AR platform. Today's high-tech GNSS receivers combine two standards (GPS, GLONASS) on two frequency bands (L1 and L2 for GPS and G1 and G2 for GLONASS).

A Novatel OEMV-1 L1 DGPS/RTK receiver with an external antenna is used, which is eliminating multi-path signals. Furthermore, a 3G modem is used for connecting to the internet to receive NTRIP (Networked Transport of RTCM via Internet Protocol) GPS correction data via a serial connection from nearby reference stations. This continuous data stream is used to achieve positional accuracies in the sub-meter to centimeter range. Figure 38 and Figure 39 show the setup developed for outdoor AR.

A seven-port USB hub connects all sensors with the USB port of the UMPC. The energy demands of these sensors together led to use a special Lithium polymer battery (3200mAh) that also supplies the GPS receiver and via a voltage transformer the USB hub.

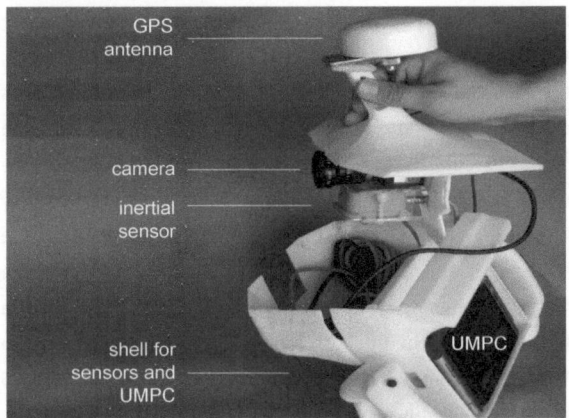

Figure 38: Pomar-3D, outdoor AR setup. Components are protected by a more robust shell.

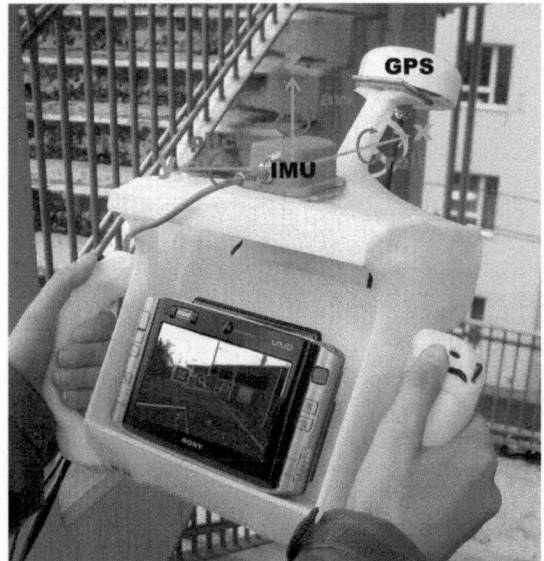

Figure 39: Pomar-3D, outdoor AR setup. outdoor AR setup. User holding the setup for visualization purposes.

Figure 40 depicts the tracking module, which has been designed towards low kinematic and high rotational movements. This is optimally suited for pedestrian outdoor user's inspecting an environment with the handheld AR device. The update rate of the tracking module consists of the of inertial sensors update rate of 25 Hz and the GPS update rate. The NTRIP GPS correction data is received every second while the GPS receiver itself has a higher update rate. For tests the EPOSA and TEPOS correction network was used (Mountpoint RTK.2-3). To use SBAS corrections if DGPS corrections are not available, the SBASCONTROL ENABLE command can be performed. The receiver automatically switches to Pseudorange Differential (RTCM or RTCA) or RTK if the appropriate corrections are received, regardless of the current setting. However that when the receiver is operating in L1 Float it is using the RTK filter. When transitioning from the RTK filter to the pseudorange filter there can be position jumps. To mitigate jumps when transitioning from/to different modes within the pseudorange filter (PSRDIFF to EGNOS correction types, for example), the PDP filter will need to be enabled using the PDPFILTER command. The system is able to perform at around 20 Hz using this configuration. Only a few mobile AR systems have been built, which are using differential GPS/RTK tracking. To the knowledge of the authors the described device is the only handheld AR system with integrated DGPS/RTK tracking and sensor fusion support.

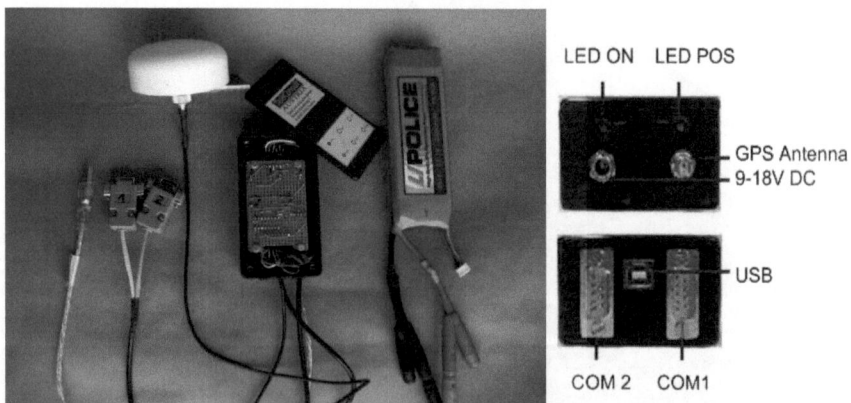

Figure 40: Differential GPS tracking module. (Left) A Novatel OEMV-1 receiver (DGPS/RTK) (11x6x3cm) is placed in the middle of connectors, antenna and battery. (Right) Side view of the receiver.

The aim of high-quality tracking and increased robustness by the robust shell around the components was reached at the prize of a clumsier setup. These observations were made during on-site tests with expert-users from companies. AR setups as described previously have the drawbacks of conventional UMPC consumer devices, e.g. no sunlight-readable display, small screen size, small batteries and insufficient ruggedness. Considering these observations, the author tried a different approach based on a rugged tablet PC. Such a computing device outperforms conventional UMPC consumer devices in the factors mentioned before.

5.6 Tablet PC-based setup

A tablet PC presents a more powerful computer compared to a UMPC. Moreover, screen size is bigger as well. Therefore, the user must accept the increased form factor of an AR setup built around a tablet PC. Figure 41 shows a mobile AR platform that is built around a rugged tablet PC (Motion J3400) with 1.6GHz Pentium CPU and sunlight viewable touch screen for real-world, field-ready outdoor conditions. The mobile platform is equipped with various sensors. A 3DoF inertial sensor (XSens MTi) — containing gyroscopes, accelerometers and magnetometers in 3D — is mounted at the encasing at the back of the AR platform. A Novatel OEMV-2 L1/L2 RTK receiver for achieving positional accuracies within the centimeter range was used.

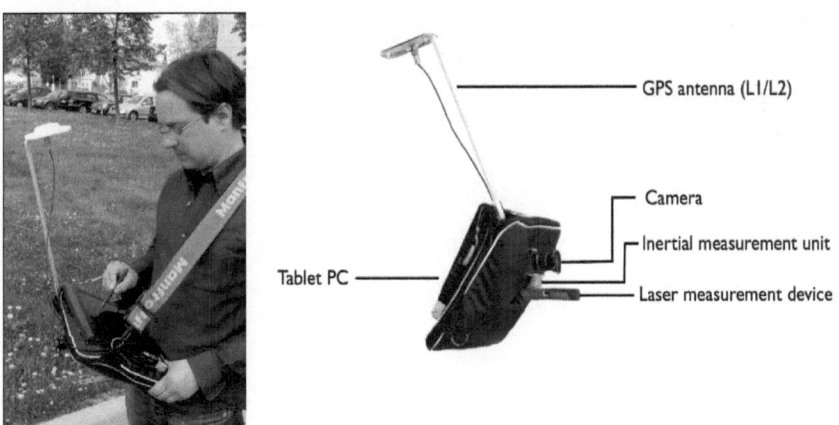

Figure 41: Tablet PC-based AR setup. The user can carry the setup using a carrying belt which allows the AR application to be controlled in a more conventional way.

The mobile AR platform can be carried using a shoulder strap or it can be mounted on a tripod. Battery power of the overall setup is enough to keep the AR application running for about three hours. This is enough for performing some tasks on-site, but not sufficient for a field worker staying on-site more than eight hours. This problem can be solved easily by just using more batteries that can be hot-swapped. Overall the tablet PC-based setup was received very well from end users. This might be the case, because the platform is superior to a UMPC platform and because tablet PCs with this form factor are already in productive use in industry. AR capability can then be reached by following a "sandwich" approach by simply mounting the sensors on the backside of the tablet PC.

5.7   Discussion

A lot of experiments have been done with various AR platforms which were designed for specific needs. Also through the many demos and on-site evaluations, the author could gain useful experience about what sensors and what kind of setup can best fulfill the expectations of users. While the Vesp'R setup received good grades for its unconventional design, usability and interaction possibilities, it is not suited for the often harsh conditions in real outdoor use. In contrast, a tablet PC-based setup can meet the requirements of outdoor use to a high degree. Although the tablet PC-based AR setup is not that fancy, user acceptance is initially high. Taking the experiences of this chapter into account, there is evidence that today's AR prototypes allow for use in demanding outdoor environments.

The three most important aspects of
debugging and real estate are the
same: Location, Location, and
Location.

R. Pattis
Carnegie Mellon University

6.   Pose tracking

Generally, 6DoF tracking of device or user pose is of central importance for augmenting the user's view with further information in a ubiquitous environment. This chapter describes several tracking approaches which seek to improve the robustness, stability and accuracy of tracking. Various methods are shown and discussed that deal with position tracking in outdoor environments. Moreover, several novel improvements for orientation tracking are presented.

6.1   Global pose estimation using multi-sensor fusion

The following tracking approaches are dedicated for outdoor environments since various GPS receivers are used for experimenting. The aim was to achieve a highly accurate position estimate in outdoor environments. Typically, GPS is used as the primary tracking system in wide-area outdoor environments. But, GPS only provides a satisfying position estimate when enough satellites are visible to the receiver and when using differential corrections from a reference station network. Moreover, magnetic orientation tracking in outdoor environments faces a number of challenges due to permanent and transient electromagnetic influences.

The overall tracking framework presented here uses Kalman filtering with a constant velocity model for fusion of DGPS/RTK with barometric heights and uses an IMU with gyroscopes, magnetometers and accelerometers to improve the transient oscillation. In the following, this approach is described in more detail. For compensation of environmental electromagnetic influences, additionally a drift-free visual tracker is applied. By online mapping of the unknown environment, this tracker allows for detecting and correcting the deviation of the 3-axis compass, which increases the robustness and accuracy of the pose estimates. Figure 42 shows the multi-sensor fusion system architecture.

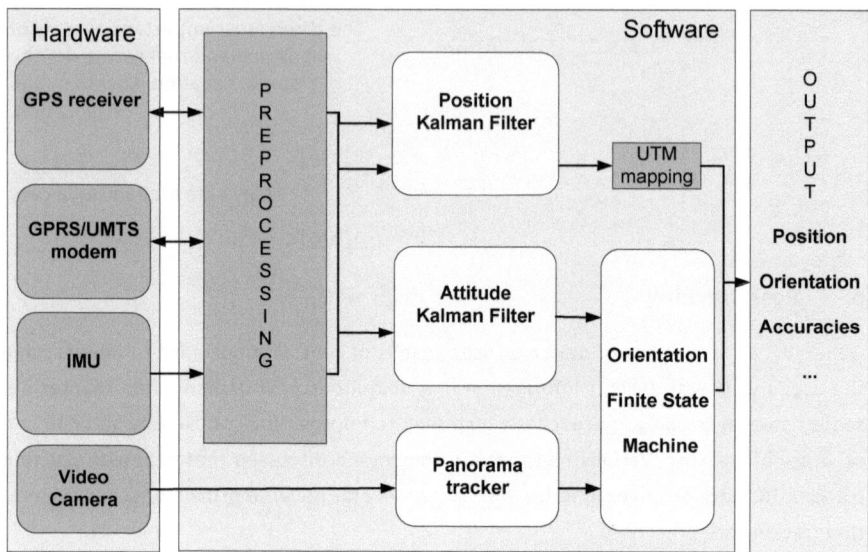

Figure 42: Multi-sensor fusion system architecture. The architecture consists of the key elements Position Kalman Filter and Orientation Finite State Machine fusing data from the Attitude Kalman Filter and the visual Panorama tracker.

Since the considered application domain shows only little kinematic motion, using GPS for supporting the orientation estimates is not useful. Considering this, position and attitude is optimized in two separate filters. A dedicated Kalman filter component for position estimation is complemented with an Attitude Kalman filter for orientation estimation. To allow for correction of both deviation and bias the visual panorama tracker is combined with the Attitude Kalman filter using a Finite State Machine. To fulfill the high requirements of the application scenario concerning positional accuracy, a DGPS/RTK receiver is employed using differential corrections from the Austrian Positioning Service ("APOS - Austrian Positioning Service", 2010). The correction data from the reference station is delivered to the handheld device in RTCM 2.3 format and thereby reduce influences such as ionospheric or tropospheric effects. This way the accuracy of the position estimation can be improved significantly.

Data transmission is done via a 3G modem connection. A special software module was developed handling the dial-in procedure, the data routing, the data conversion and the data transfer to the GPS receiver. If the 3G connection is lost, the software module reconnects automatically. Moreover, lever-arm correction is performed, the data is prepared and the

GPS receiver uses this data for calculating a DGPS/RTK position estimate. Next, the Position Kalman filter fuses the position estimate with the barometric height for the final position estimate, which is transformed into Universal Transverse Mercator (UTM) format.

Raw data of the accelerometers, gyroscopes and magnetometers are preprocessed and converted. Then the Attitude Kalman filter fuses the delivered data resulting in roll, pitch and yaw as output. Two effects occur in combination with the attitude. First, the magnetic yaw is afflicted with a deviation. Second, the angular velocities of the gyroscopes show a bias, which results in a drift of the angles. The bias of the gyroscopes is considered and corrected by the filter estimation.

The magnetic yaw is deducted from the 3-axis magnetometer and refers to compass north. While the variation can be corrected, the deviation effect represents an unknown location-dependent term. Magnetic yaw and the angular rate of yaw of the gyroscope should support each other. Both have a variable term namely deviation and bias. Consequently, the simultaneous estimation of the deviation and bias of yaw cannot be determined within the Attitude Kalman filter.

Only an additional input for yaw, which is without drift and without bias, would allow solving this problem. To solve this problem and to estimate the deviation of the magnetometer yaw, a visual panorama tracker as additional input is introduced and applied. By online tracking natural features and simultaneously mapping the environment, the visual tracker delivers drift-free and unbiased orientation estimates. At the beginning, the visual panorama tracker is initialized with the pitch of the inertial tracker, so that the user does not need to hold the AR device horizontally at start up. Furthermore, the visual panorama tracker is able to use the motion model of the inertial sensor to provide more accurate priors under fast motions. Classically, inertial sensors are better suited for measuring high-frequency and rapid motion while the slower vision sensor performs best with low frequency motion and provides absolute references to reset the error. This multi-sensor fusion approach allows for detection and correction of both drift and bias of the inertial sensor.

6.2   Position tracking

To obtain the required position accuracy in real-time, GPS has to be used either in differential positioning mode (DGPS) or in real-time kinematics (RTK) positioning mode. DGPS systems improve accuracy by broadcasting correction information from a stationary base station to roving users, based on comparing the computed position with the known position of a carefully surveyed reference antenna. RTK GPS uses information about the GPS signal's carrier phase at the base station and the rover to reach even better (centimeter-level) accuracy. GPS is line-of-sight and it loses track easily when indoors, under tree cover, or near tall buildings (in particular in so called "urban canyons"). GPS signal loss is often addressed through dead-reckoning techniques that rely on tetherless local sensors, such as magnetometers, gyroscopes, accelerometers, odometers, and pedometers (Hallaway, 2004). The outdoor RTK GPS+GLONASS system has a maximum tracking resolution of 1–2 centimeters at an update rate of up to 1–2 Hz. Its accuracy may degrade to meter-level when fewer than four satellites are visible. If communication to the RTK base station is lost, a fall back to an uncorrected accuracy of 2-3 m using EGNOS (European Geostationary Navigation Overlay Service) occurs. EGNOS is a satellite based augmentation system (SBAS) under development by the European Space Agency, the European Commission and EUROCONTROL. It is intended to supplement the GPS, GLONASS and Galileo systems by reporting on the reliability and accuracy of the signals.

Due to the measurement principle, the vertical GPS accuracy (height) is in general two times worse than the horizontal GPS accuracy. After initializing the barometer (using GPS heights as reference), the barometric height is more stable than GPS heights in particular in the RTK mode. Additional position information helps in overcoming GPS shadowing in urban regions. Therefore, the Position Kalman filter performs a sensor fusion between GPS and barometer. The GPS height and the barometric height are combined in the filter with respect to their accuracies. Accurate measurements get higher weights. During the initialization step, the barometric height gets a small weight in such a way that the filtered height is exclusively affected by the GPS height. The equations of the Kalman filter are well-known and are therefore not repeated here (Hofmann-Wellenhof, Legat, & Wieser, 2004). The filter input consists of:

- GPS position (geographical coordinates)
- Barometric height (offset corrected)

The **observation vector** z and the state vector x are defined as followed.

$$z = \begin{bmatrix} \phi_{GPS} & \lambda_{GPS} & h_{GPS} & h_{BARO} \end{bmatrix}^T$$
$$x = \begin{bmatrix} \phi & \lambda & h & \dot{\phi} & \dot{\lambda} & \dot{h} \end{bmatrix}^T \qquad (1)$$

The barometric height is used as absolute height after correcting for the barometric off-set. The barometric offset is estimated at the beginning. During the offset estimation the barometric height is not integrated in the filtering.

The **observation equations** are linear and stated below.

$$\phi_{GPS} = \phi$$
$$\lambda_{GPS} = \lambda$$
$$h_{GPS} = h$$
$$h_{BARO} = h$$

(2)

The application which is covered in this paper implicates little dynamics. Therefore it is sufficient to consider a uniform motion for the dynamic model. In addition to the geographical coordinates itself, also their velocities have to be estimated in order to perform prediction. The dynamic model simply updates the position through integrating the velocities.

6.3 Orientation tracking

The outdoors presents enormous challenges for mixed and augmented reality. Outdoor environments encompass extreme weather and illumination conditions, and mobile systems must deal with technological constraints, including low-resolution cameras and displays, inaccurate and fragile tracking systems, limited system and network resources, and cumbersome interaction devices.

The Attitude Kalman filter performs a sensor fusion of gyroscopes, accelerometers and magnetometer. Pre-processed quantities of these sensors form the filter input and are described in detail later on:

- Roll and pitch derived from triaxial accelerometers
- Magnetic yaw
- Gyroscopic angular rates of roll, pitch and yaw

The **observation vector** z and the state vector x are defined in the following way.

$$z = \begin{bmatrix} \varphi_{ACC} & \vartheta_{ACC} & \psi_{MAG} & \dot{\varphi}_{GYR} & \dot{\vartheta}_{GYR} & \dot{\psi}_{GYR} \end{bmatrix}^T$$

$$x = \begin{bmatrix} \varphi & \vartheta & \psi & \dot{\varphi} & \dot{\vartheta} & \dot{\psi} & \ddot{\varphi} & \ddot{\vartheta} & \ddot{\psi} & b_\varphi & b_\vartheta & b_\psi \end{bmatrix}^T \qquad (3)$$

Gyroscopic angular rates have biases which are estimated in this filter to avoid a temporal drift of the attitude angles. The linear **observation equations** are the following.

$$
\begin{aligned}
\varphi_{ACC} &= \varphi \\
\vartheta_{ACC} &= \vartheta \quad (4) \\
\psi_{MAG} &= \psi
\end{aligned}
\qquad\qquad
\begin{aligned}
\dot{\varphi}_{GYR} &= \dot{\varphi} + b_\varphi \\
\dot{\vartheta}_{GYR} &= \dot{\vartheta} + b_\vartheta \\
\dot{\psi}_{GYR} &= \dot{\psi} + b_\psi
\end{aligned}
\qquad (5)
$$

For a sufficiently small time interval in accordance with the measurement update rate (here 25 Hz), the biases are assumed to be constant. A uniform angular acceleration of the attitude angles is assumed for the dynamic model. Therefore, the angular rates and the angular accelerations of the attitude angles have to be estimated within the filter. The filtered magnetic yaw is still affected by magnetic deviation and magnetic declination.

6.3.1 Gyroscope measurement model

A triaxial gyroscope measures raw angular rates along the input axis in the body frame (BF).

$$\omega^{BF} = \begin{bmatrix} \omega_x^{BF} & \omega_y^{BF} & \omega_z^{BF} \end{bmatrix}^T \qquad (6)$$

These measurements can be converted into angular rates of roll, pitch and yaw. The angular rate of roll corresponds directly to the measured angular rate along the x-axis.

$$\dot{\varphi} = \dot{\varphi}\left(\omega_x^{BF}\right) = \omega_x^{BF} \qquad (7)$$

The angular rate of pitch is not only a function of measured angular rates along the y- and z-axis but also a function of the roll angle. It can be derived as followed.

$$\dot{\vartheta} = \dot{\vartheta}\left(\omega_y^{BF}, \omega_z^{BF}, \varphi\right) = \omega_y^{BF} \cdot \cos\varphi - \omega_z^{BF} \cdot \sin\varphi \qquad (8)$$

The angular rate of yaw is a function of all the raw angular rates of the tripod and the leveling angles roll and pitch.

$$\dot{\psi} = \dot{\psi}\left(\omega_x^{BF}, \omega_y^{BF}, \omega_z^{BF}, \varphi, \vartheta\right) = -\omega_x^{BF} \cdot \sin\vartheta$$
$$+ \omega_y^{BF} \cdot \sin\varphi\cos\vartheta + \omega_z^{BF} \cdot \cos\varphi\cos\vartheta \tag{9}$$

6.3.2 Magnetometer measurement model

A triaxial magnetometer is needed to derive heading information. Thereby magnetic field strengths along three input axis in the body frame are measured.

$$m^{BF} = \left[m_x^{BF} \quad m_y^{BF} \quad m_z^{BF} \right]^T \tag{10}$$

In this context, the heading information derived from the magnetometer is called magnetic yaw (MY). The magnetic yaw does not equal the true yaw (TY). Due to the fact that a magnetometer corresponds to compass north, the author prefers to stay with the predicate 'magnetic'. The magnetic yaw is infected by magnetic variation (VAR) as well as magnetic deviation (DEV).

$$TY = MY + DEV + VAR \tag{11}$$

The magnetic yaw can be derived, if the leveling angles roll and pitch are known according to (Caruso, 1998).

$$\psi = \psi\left(m_x^{BF}, m_y^{BF}, m_z^{BF}, \varphi, \vartheta\right) = \tan^{-1}\left(\frac{-H_y}{H_x}\right)$$
$$H_x = m_x^{BF} \cdot \cos\vartheta + m_y^{BF} \cdot \sin\vartheta\sin\varphi + m_z^{BF} \cdot \sin\vartheta\cos\varphi \tag{12}$$
$$H_y = m_y^{BF} \cdot \cos\varphi - m_z^{BF} \cdot \sin\varphi$$

6.3.3 Accelerometer measurement model

The principle to obtain roll and pitch from a triaxial accelerometer according to Groves (Groves, 2008) is called leveling. This method implies that the observed accelerations along the input axis in the body frame are exclusively due to the gravitational acceleration.

$$f^{BF} = \left[f_x^{BF} \quad f_y^{BF} \quad f_z^{BF} \right]^T \tag{13}$$

The sensor must not be affected by additional accelerations. That accounts static measurements. With regard to this application the condition for using this method will be partly fulfilled. In case of sensor movement, the derived leveling angles get lower weight for the filtering. The leveling equations are stated below.

$$\varphi = \varphi\left(f_y^{BF}, f_z^{BF}\right) = \tan^{-1}\left(\frac{f_y^{BF}}{f_z^{BF}}\right) \tag{14}$$

$$\vartheta = \vartheta\left(f_x^{BF}, f_y^{BF}, f_z^{BF}\right) = \tan^{-1}\left(\frac{-f_x^{BF}}{\sqrt{f_y^{BF^2} + f_z^{BF^2}}}\right) \tag{15}$$

6.4 Visual tracking approach

This subsection briefly describes the implementation of the visual panorama tracker (Wagner, Mulloni, Langlotz, & Schmalstieg, 2010) used for visual tracking. For compensation of the deviation of the inertial sensor, which is induced by electromagnetic influences, additionally this visual tracker for detecting and correcting the deviation of the 3-axis magnetic compass is applied. This visual tracker improves both accuracy and robustness of the rotation estimation. The visual panorama tracker assumes a purely rotational motion, ignoring any translational movement. A pure rotation does not create a parallax effect and hence the environment can be mapped onto a closed two-dimensional surface, such as a cube, sphere or cylinder. This technique is well known in computer graphics under the names environment mapping, reflection mapping or sky-box. The visual tracker maps the environment onto a cylinder with a height of Π/2 relative to the radius, and can therefore map ~76.3 degrees vertically. Conceptually, the radius does not matter, so for convenience it is set to 1.

Starting with a predefined initial direction, the tracker maps the environment on the fly, while the camera is moving. A similar technique has been presented by (DiVerdi, Wither, & Höllerer, 2008). However the mapping and tracking technique is much more efficient: DiVerdi's approach requires intensive GPU processing to run in real-time, whereas this approach runs in real-time with minimal memory and CPU resources only. This is a significant characteristic, since makerless systems generally suffer from high computational costs. Compared to high-end PCs, small handheld devices, such as UMPCs have slow CPUs and GPUs. Hence, an efficient solution is mandatory. The tests showed that the visual tracker requires between 1.5ms and 2.5ms per frame on a notebook with a 2.5GHz CPU and between 4.0ms and 6.5ms per frame on the handheld device. The speed depends on the number of new pixels that are drawn into the map. For a completed map the tracker therefore runs in ~4.0ms per frame on the UMPC, leaving enough processing power for other tasks.

Even though its technical details are entirely different from traditional approaches of simultaneous localization and mapping (SLAM), the basic approach is similar: For each frame, the tracker first estimates a new pose from the camera image and then enters new features into the map. A major difference to classic SLAM systems is that this approach creates a dense map, but entries are not updated, once they are mapped. This is viable, because under a pure rotational motion the whole 2D state of a map feature is directly observable and the motion is sufficiently constrained. The orientation update step uses 2D-2D point correspondences between the environment map and the camera image. The point correspondences are matched using normalized cross correlation (NCC) on warped 8x8 pixel patches. The locations of the interest points are selected in the map using the FAST corner detector in areas of the map that have already been finished. These keypoint locations are then projected into the camera image and searched in the close proximity to yield sub-pixel accuracy. It is important to notice that no keypoint detector is applied, such as FAST on the whole camera image, which is one reason for the high speed of this approach.

Once enough correspondences have been found (usually the tracker finds at least a several hundred), the tracker updates the rotation using Gauss-Newton iteration: Basically the same optimization is performed as for a full camera pose, but the position is ignored and the Jacobians are only calculated for the three rotation parameters. An M-estimator is used to deal with re-projection errors. The final 3x3 system is then solved using Cholesky decomposition to yield the update vector for the rotation. Since the starting point is already close to the final solution only few (3-5) iterations are required. After the rotation has been updated to match the current camera image, the camera image is projected into the environment map. Run-length encoded pixel spans are used to keep track of which parts of the map have already been mapped and which haven't. Hence, every pixel of the map is filled only once. When a pixel in the map has been selected to be filled, its 3D cylinder position is intersected with the camera image, the image coordinate is undistorted, bilinear filtering is used to extract the pixel intensity and finally for correcting vignetting a simple radial falloff model is used.

The visual tracker works at a camera resolution of 320x240 pixels. The map is created at a resolution of 2048x512 pixels. For a typical camera field of view of 60 degrees, the camera's resolution is therefore close to the map's resolution: 320 pixels / 60 · 360 = 1920 ≈ 2048 pixels. The angular resolution of the map is therefore 360 degrees / 2048 pixels = 0.176 degrees per pixel with an average error of ~0.002 degrees per pixel (see Section 5.3). This is much more accurate than a gyroscope sensor can deliver. Although errors accumulate over the map, the tracker is inherently free of drift. The visual tracker can perform loop closing to remove the accumulated error. The tracker therefore uses a map that is larger than 360 degrees horizontally (e.g. 405 degrees) in order to create overlapping regions. Once enough overlap is available, the tracker uses template matching on the keypoints in the map to identify the exact overlap. Since no error model is available, the map is simply scaled and sheared to close the gap.

6.5   Fusion of attitude with visual tracking

Now, having two variants for 3DoF orientation tracking, namely the Attitude Kalman Filter and the visual based tracker, the question was how to combine both in a meaningful way. The author has implemented the integration of the inertial sensors attitude with the visual tracker orientation output using a state machine. This state machine is denoted as SM (Σ, S, S_0, δ) and shown in Figure 43. Σ describes the five input conditions, S is the set of four states, S_0 is the initial state, an element of S, δ is the state-transition function: $\delta : Sx\Sigma \rightarrow S$. Table 2 lists the combinations of states and transition conditions. ΔP (delta pose) denotes the difference between the orientations of visual and inertial tracker. The state machine starts in state S_0, in which both the inertial and the visual tracker are valid.

Despite a fix offset between the orientations of both trackers, the deviation is zero. The visual tracker uses the motion model of the inertial tracker to provide more accurate priors under fast motions. The final orientation is calculated by fusing the orientation of the visual tracker with the orientation of the inertial tracker. Timestamps assure a correct synchronization of both trackers.

A threshold set on ΔP is responsible for deciding whether the two trackers are diverging or not. This threshold is empirically determined and is less than a few degrees. A second threshold, tracking the number of detected features in the image (20 in these experiments), is used for assessing the validity of the visual tracker. If more features than the threshold have been found in the image, the visual tracker can be trusted.

In case magnetic deviation occurs, the orientation of the inertial tracker will change in respect to the orientation of the visual tracker. As long as the deviation varies, the visual tracker, which is now using its constant velocity motion model, is trusted more and its tracking results are taken as final orientation result (S1).

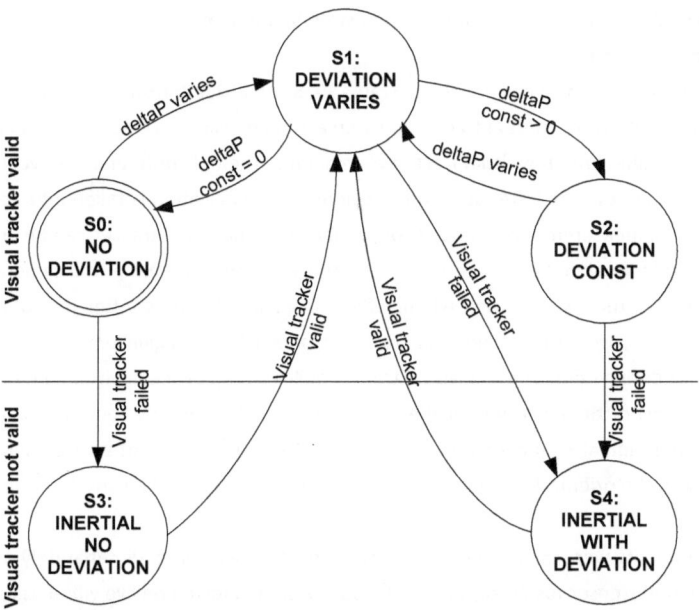

Figure 43: State machine SM {Σ, S_0, S1, S2, S3, S4, δ}. Transitions are made depending on the inertial sensors attitude and the visual tracker orientation output.

Table 2: State transition table. Shows which conditions induce a transition to which state.

Condition (δ)	Current State (S)				
	S0	S1	S2	S3	S4
ΔP varies	S1	S1	S1	-	-
ΔP const = 0	S0	S0	-	-	-
ΔP const > 0	-	S2	S2	-	-
Visual tracker failed	S3	S4	S3	S4	S4
Visual tracker valid (re-initialized)	-	-	-	S1	S1

In case the deviation decreases back towards zero, the transition is made to S_0 (e.g. if a transient deviation occurs such as a large vehicle driving by).

In case the introduced deviation stays constant and bigger than zero, the state machine changes to state S2. In this state also the visual tracker is trusted, since the inertial tracker is precise, but not accurate now (e.g. a constant deviation can be induced if the user moves into an area with different magnetic fields or if a magnetic field is introduced around a static user,

for example, by switching on electric circuits). Also in state S2 the motion model of the inertial tracker is used for vision tracking.

If the visual tracker fails, which can happen in states S0, S1 and S2, a transition to state S3
or S4 is performed, in which the inertial tracker must be assumed valid, since else no further
tracking would be possible until the visual tracker re-initializes. The only difference between
the state S3 and S4 is that in the latter state the previous known deviation is taken into account with the inertial measurements. Reasons for the failing of the visual tracker can be an
abrupt or very fast rotational motion of the AR device. While the video stream of the camera
delivers blurred images, the visual tracker will not find meaningful features. After the rotational motion decreases, the visual panorama tracker cannot be used again until it is re-
initialized. Note that during operation, the visual tracker builds a map of the environment on
the fly. In case the part of the environment the camera is facing has already been mapped,
the tracker is able to re-initialize immediately to the correct pose. If the current environment
is new to the tracker, the orientation values of the inertial tracker need to be considered for
re-initialization.

Extensive experiments were conducted to test the single tracking solutions separately as
well as the overall multi-sensor fusion approach. The approach was tested on live video using
the hardware presented in Chapter 5.6 (see Figure 38). A test site near the campus was chosen. For this experiment at the test site an urban 3D model from data using geospatial databases was generated. The data was delivered by the local utility company as described in
(Schall & Schmalstieg, 2008).

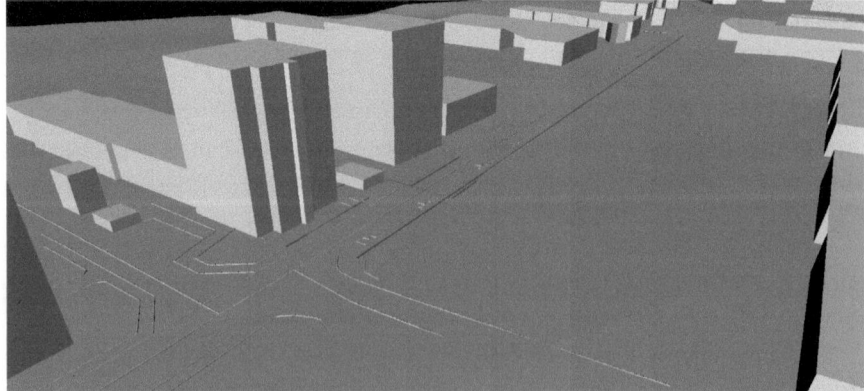

Figure 44: 3D model of the test site. The model includes a DEM, extruded building footprints and
underground water and electricity lines.

The model (see Figure 44) includes a DEM of the test site, extruded building footprints, water and electricity lines, pavement border lines, street middle axis as well as surveyed reference points that can act as ground truth data.

6.6   Position using differential GPS

This test series uses the Position Kalman filter, which integrates the GPS and barometer sensor. A first test assessed the positional accuracy and precision of the GPS receiver and the APOS service in a static scenario. The comparison in Figure 45 shows the GPS C/A Code solution and the DGPS solution. Qualitatively it is visible that a more stable position can be calculated by using correction signals. Table 3 lists the measurements according to the plots in Figure 45. Furthermore, a test of the Position Kalman filter was performed in a dynamic scenario, in which the user moves along a path and passes over known reference points, which are drawn in red color (see Figure 46). The test shows that the filtered DGPS position (in blue color) satisfies the accuracy requirements. After one minute, instead of the C/A Code solution, a DGPS position solution with higher accuracy is calculated by the receiver. After five minutes, positions estimates with sub-meter accuracy are calculated by the static receiver. Then, the known reference points were passed over with sub-meter accuracy. It can be observed that when the user changes the direction of movement, a short post-pulse oscillation appears (see grey ellipsoids).

Table 3: GPS accuracy measurements using GPS and DGPS.

	GPS C/A code		DGPS	
	Mean [m]	Stddev [m]	Mean [m]	Stddev [m]
X	1.058	2.600	0.800	1.848
Y	0.617	1.296	0.765	1.553

Table 4: GPS accuracy measurements of the L1/L2 differential GPS receiver.

	Northing [m]	Easting [m]	Height [m]
Reference point	214300,680	-66811,806	378,127
Mean	214300,733	-66811,872	378,320
Difference	0,052	-0,065	0,132

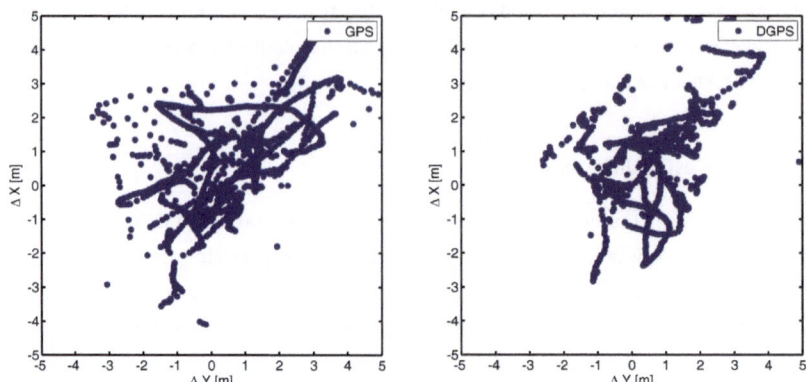

Figure 45: GPS measurements. GPS C/A code solution (left). DGPS solution (right).

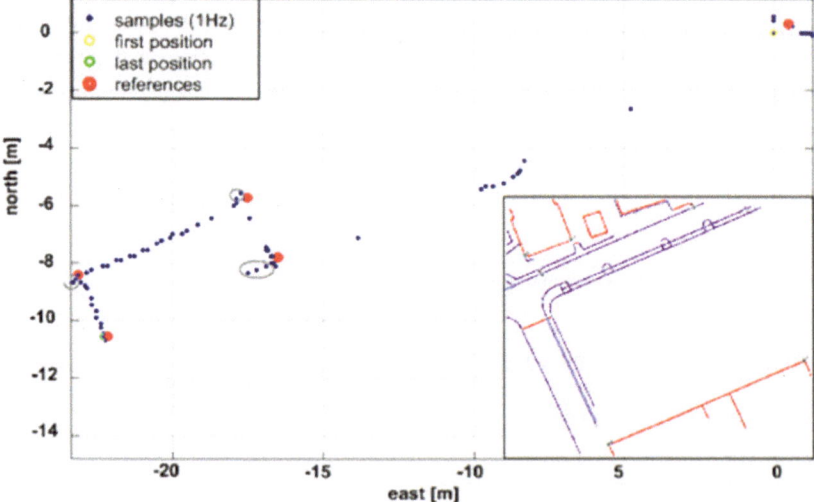

Figure 46: Position estimates along a path using the Position Kalman filter. Gray ellipsoids depict post-pulse oscillation of the filter. Inlay: view of the cadastral map of the test site. Blue lines are vector map features of streets and pavements. Violet lines show building footprints with reference points at their corners.

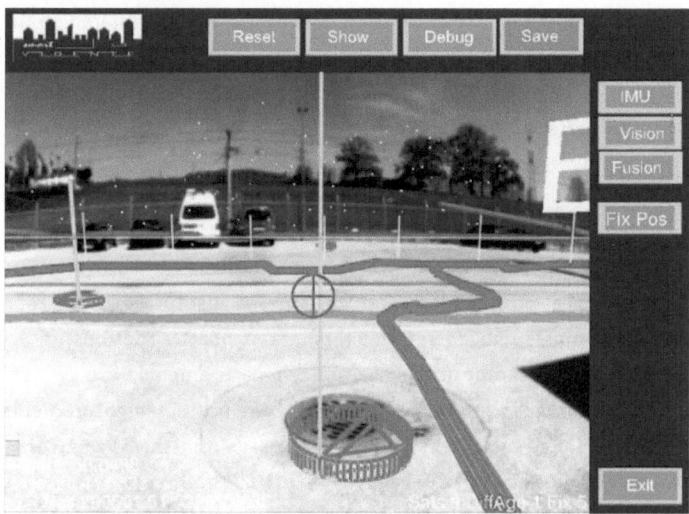

Figure 47: AR overlay. (Top) AR view of superimposed circular capping using the L1 differential GPS receiver at a test site in Salzburg. (Bottom) AR view of superimposed rectangular capping using the L1 differential GPS receiver at a test site in Salzburg.

Figure 46 shows examples how well virtual cappings (in blue color) are registered on the real environment using the differential GPS receiver. Electricity lines are rendered in red color and reference points are drawn as vertical lines. The accuracy of the GPS is good, but not satisfying the desired accuracy better than 10 cm.

6.7 Position using real-time kinematic GPS

A series of measurements with the differential L1/L2 GPS receiver from Novatel was performed to evaluate a typical positional accuracy in 3D. The Institute of Navigation and Satellite geodesy maintains surveyed reference points at nearby rooftops at the campus. Figure 48 shows the test setup for the accuracy measurements. The GPS antenna was exactly placed at surveyed reference points while the position measurements were performed. The differential correction signals were used from the EPOSA reference system ("EPOSA - Echtzeit Positionierung Austria", 2010). The graphs depicted in Figure 49 shows results of a representative position measurement at a surveyed reference point at the roof using the differential L1/L2 GPS receiver. The mean error is in the range from 5 to 6 centimeters for northing and easting and 13 centimeters for height. Using this differential GPS receiver in the AR setup it can be assumed to achieve positional accuracies with an error of 10 centimeters or below.

The previous experiment evaluated the accuracy of the L1/L2 GPS receiver only. But, the overall registration error in 3D includes next to the position inaccuracy of the GPS also the orientation errors of the IMU.

 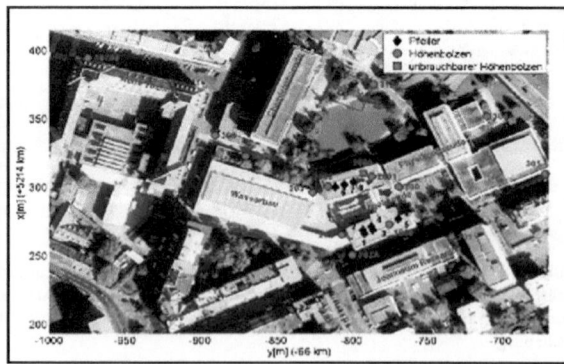

Figure 48: Rooftop test scenario. (Left) Rooftop test setup for accuracy measurements. GPS antenna is placed at surveyed reference points while performing position measurements. (Right) Surveyed reference points at the rooftops visible from the Institute of Navigation and Satellite Geodesy.

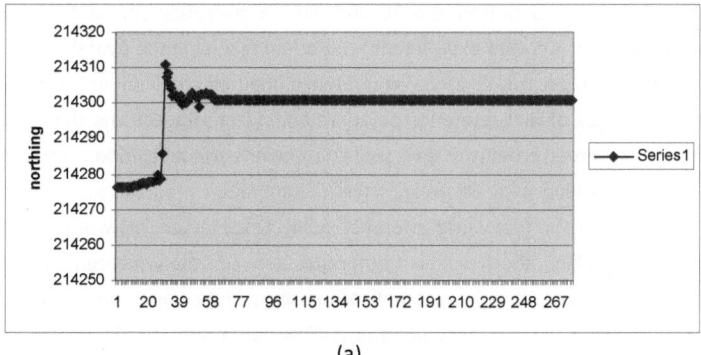

(a)

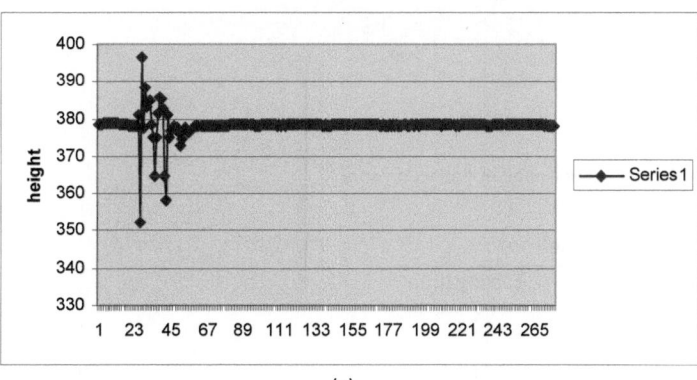

(b)

(c)

Figure 49: GPS accuracy measurements of (a) northing, (b) easting and (c) height using the L1/L2 differential GPS receiver.

Furthermore, inaccurate base data from the GIS and camera calibration add to the registration (or re-projection) error. Another experiment focused on assessing the overall re-projection error. Therefore a highly accurate surveyed reference point on a pavement is used as ground truth. The coordinates of that reference point are taken from the GIS and the reference point is visualized as a green cross with a vertical line. Under perfect conditions (perfect position and orientation tracking with no errors, camera calibration etc.) the green cross would exactly be visualized at the real-world reference point. Since under real-world conditions, there is no perfect tracking, there will be a difference between the visualized and the real-world reference point. Figure 51 shows example screenshots of the superimposed reference point. The coordinates of the reference point in WGS84 datum are 15° 27' 25,36953'', 47° 03' 31,76561'' and 357.776 m height. The equivalent coordinates in Gauß-Krüger coordinates (M34) are -66489,050 easting and 213671.953 northing. A grid with one centimeter square distance has been put on the pavement (see Appendix 9.2, Figure 85).

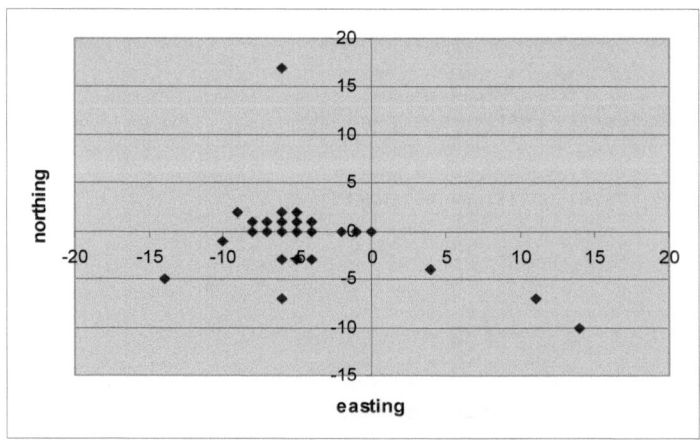

Figure 50: Error distribution of the overall re-projection error.

Table 5: Error in the overall re-projection.

	Northing [m]	Easting [m]
Mean	0,00	-4,78
Std dev	3,20	3,34

Figure 51: AR view of a re-projected physical reference point on the pavement. (Top) User looks eastward. (Bottom) User looks westward.

Furthermore, more circular rings have been plotted starting with a radius of 5 centimeters for the inner circle and increasing the radius by 5 cm for each outer ring. A small hole was cut out from the middle of the plotted grid and exactly placed on the real-world reference point. The re-projection error of the reference point is around 5 centimeters.

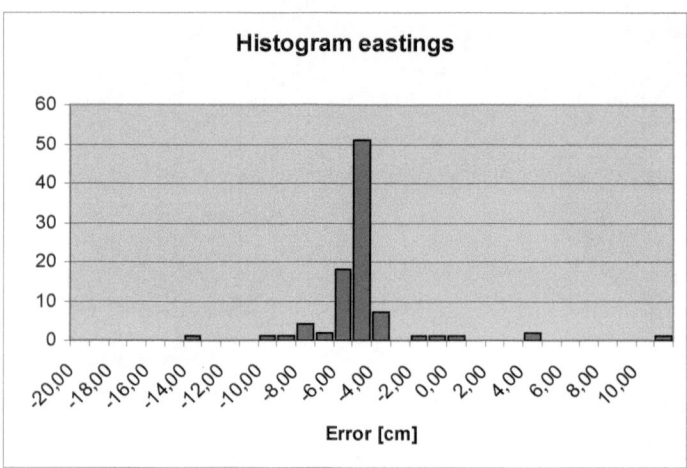

Figure 52: Histogram of re-projection errors for easting direction.

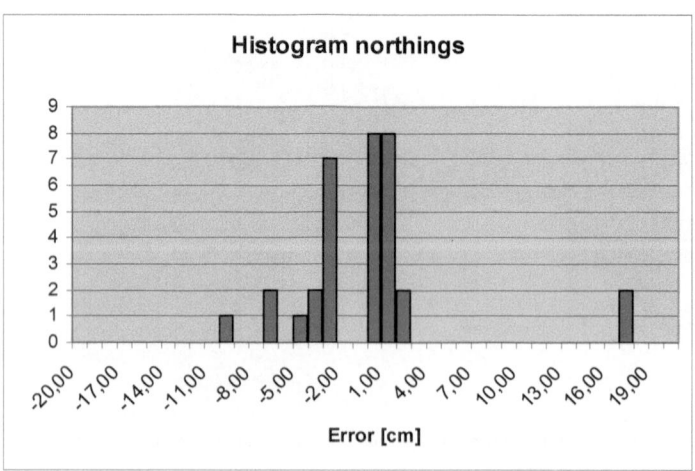

Figure 53: Histogram of re-projection errors for northing direction.

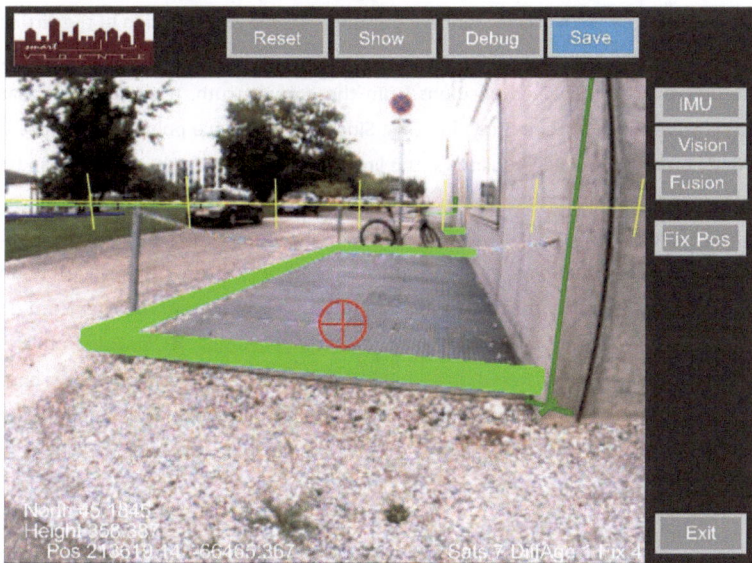

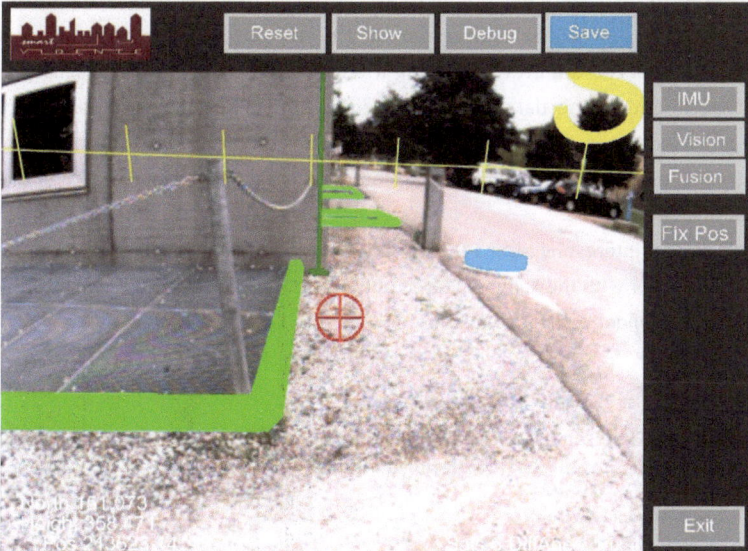

Figure 54: AR view with superimposed enclosures and base point of the building corner and a capping registered in 3D. (Top) user looking in eastwards. (Bottom) user looking in southwards.

Figure 50 shows the mean error and standard deviation for the re-projection of the virtual reference point at the real-world reference point as recorded in the experiments. Figure 51 illustrates the offset of the observed positions from the ground truth. It can be seen that there is a shift westwards of around 5 centimeters. Since the reference point is at the pavement and already quite close to high buildings, the line of sight to some GPS satellites can be lost. This may lead to a less accurate position estimate.

Figure 52 and Figure 53 depict the histograms of the re-projection error in both directions. It can be seen that the distribution for northing direction is more spread than the distribution for easting direction. Furthermore, for having a better impression of the registration accuracy, Figure 54 illustrates AR views with superimposed geospatial 3D models from the GIS. The geospatial models are registered in 3D and include features such as enclosures, base points of the building corner and cappings. The red circle with the cross in the middle of the screen is an aid for interacting with objects.

6.8   Attitude Kalman filter

Moreover, a series of tests were performed to assess the accuracy of the inertial sensor as well as the stability and behavior over time of the Attitude Kalman filter. One test was conducted to observe the attitude during a test of the relative angular accuracy. During this experiment the inertial sensor was fixed on mount that was rotated in steps of 90 degrees between the five measurements of yaw. The duration of one measurement was 15 minutes. Table 6 shows that the relative accuracy of yaw is better than one degree under ideal conditions.

Next, a selection of interesting results from a series of measurements comparing the output of the inertial sensor with the Kalman filtered output is presented. In the following scenario, a user is observed under realistic conditions.

Table 6: Attitude during test of relative angular accuracy.

Magnetic yaw [deg]			Measured rotation of magnetic yaw [deg]
Deg	Mean	Std dev	
180	178.44	0.12	0.0
270	269.12	0.09	90.68
360	359.29	0.07	90.17
90	89.08	0.04	89.79
180	178.52	0.17	89.44

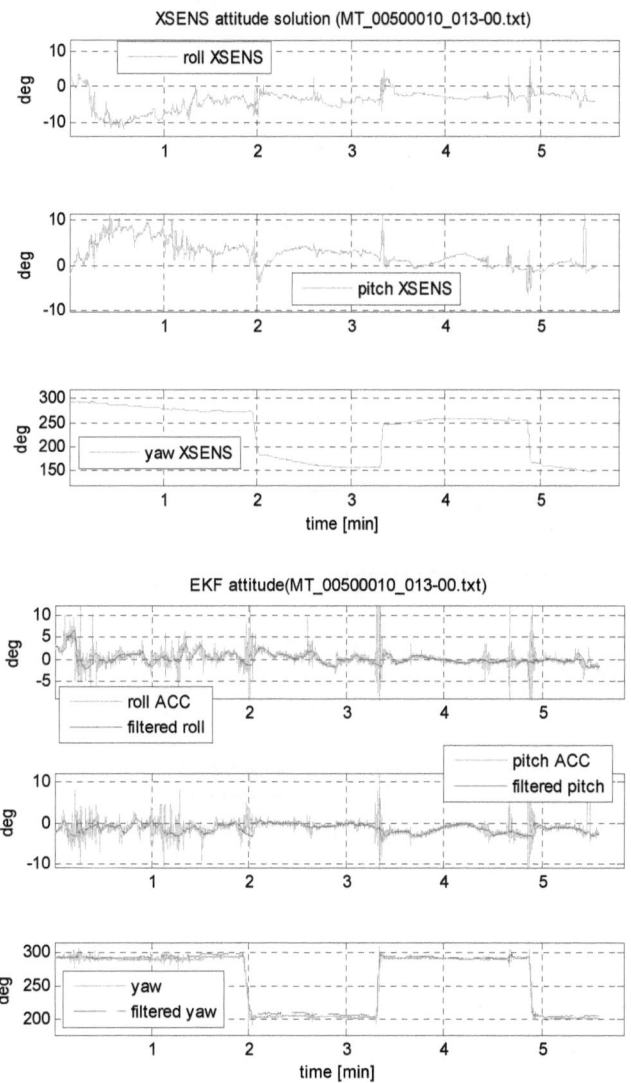

Figure 55: Attitude during situation (a): Rotation of user, no drift (up: inertial sensor, down: Kalman filter attitude).

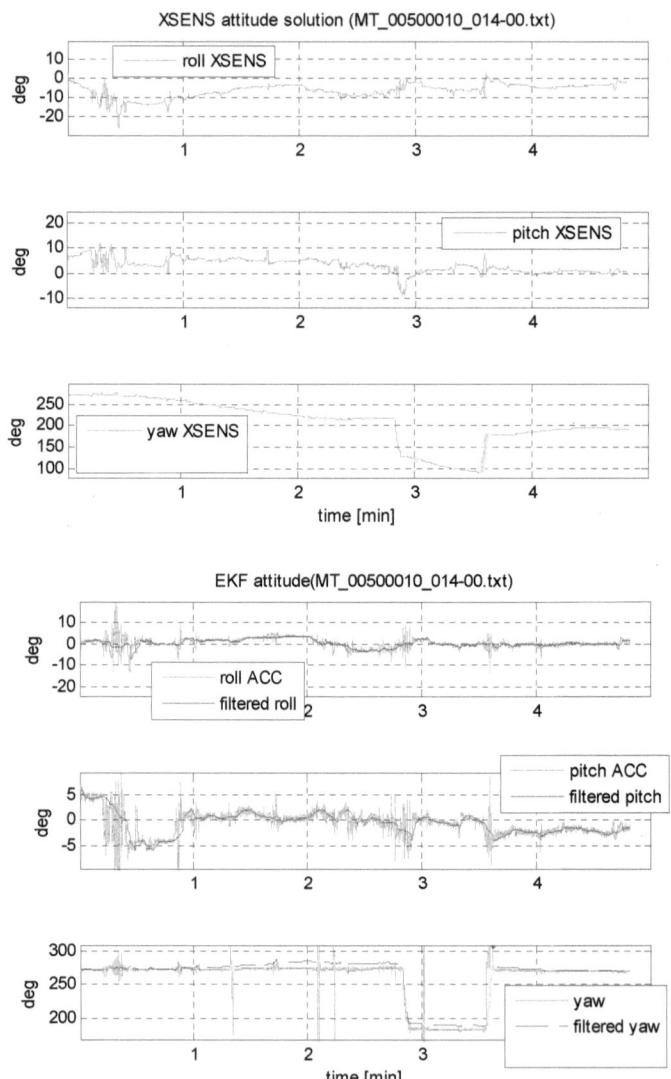

Figure 56: Attitude during situation (b): Rotation of user, transient drift (up: inertial sensor, down: Kalman filter attitude).

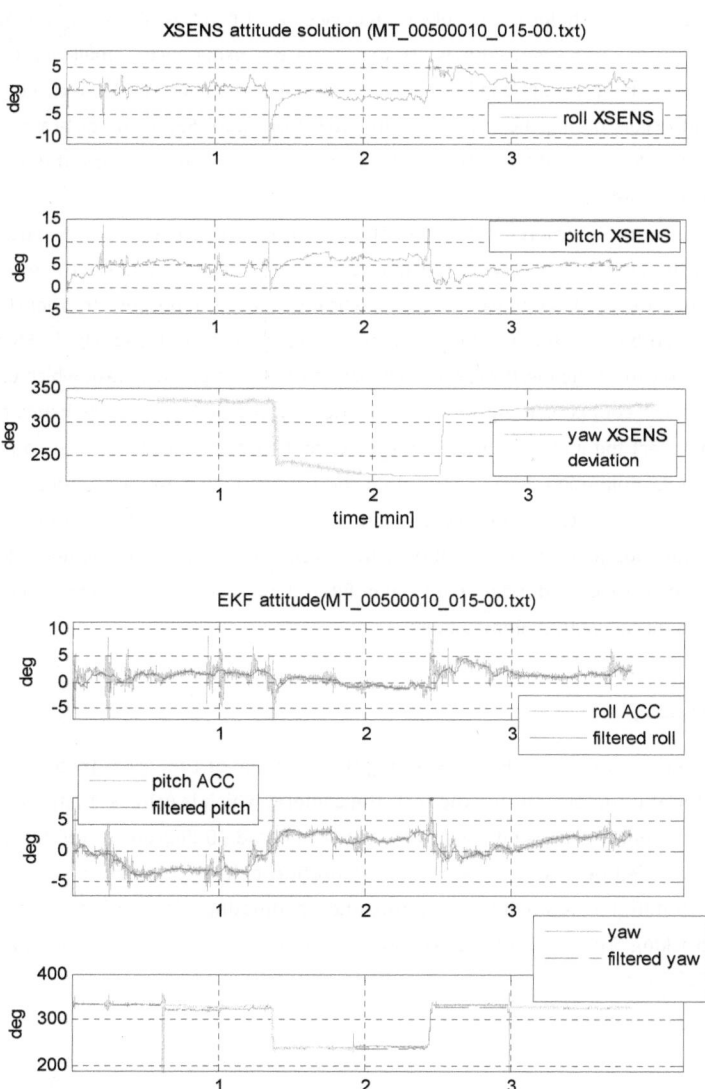

Figure 57: Attitude during situation (c): Rotation of user, permanent drift (left: inertial sensor, right: Kalman filter attitude).

In situation (a), the user holds the handheld AR device in her hands while taking the measurements. The figures show a dataset in which the sensor was experiencing small vibrations from holding it in the hands. Additionally, at a later time the user turned the AR device by 90 degrees. No drift occurs during the test shown in dataset of Figure 55. In the diagrams the drift is marked by bold dots and lines. The results of situation (b) in which transient drift occurs, are shown in Figure 56.

Figure 57 depicts the measurements of situation (c) in which permanent drift appears. The accuracy of the filtered yaw stays better than 10 degrees in all 3 situations. The experiments showed that the accuracy of yaw can be significantly increased because transient and permanent drift can be detected and their influence on yaw can be reduced. Figure 58 depicts roll, pitch and yaw of the inertial sensor that was assessed in a practical test, which used identical data for the comparison. Again, the AR platform was held in the user's hands and shivery motions affect the inertial sensor. In this practical test, the AR platform was rotated by 90 degrees in yaw direction. Results show that the Kalman filter attitude is more stable than the inertial sensor attitude. In particular, at the beginning a large difference between the attitudes occurs due to a slow transient oscillation. According to the manufacturer a lead time of 15 minutes is suggested. Using the Kalman filtered approach, no lead time is necessary and the attitude can be used immediately.

6.9   Visual tracking

Table 7 lists results from experiments assessing the accuracy of the visual tracker after having calibrated the camera. During the test, the camera, which is using a 4.2mm wide-angle lens, is fixed on a mount and rotated in steps of 30 and 90 degrees to the left and right starting at 0 degrees. At each step the orientation delivered by the tracker was measured for 15 minutes. A small bias towards underestimating the rotation is present in the visual tracking. This is corrected as soon as a loop-closure happens in the panorama.

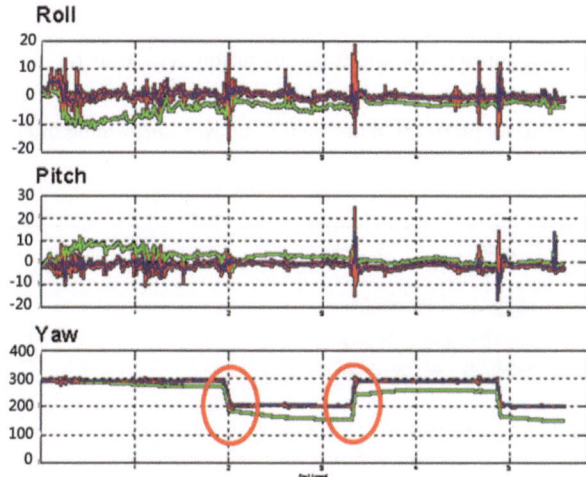

Figure 58: Comparison of orientation estimates during start up phase and rotation of inertial sensor (yaw). Built-in filter in green color. Proposed Kalman Filter in blue color.

Figure 59: Screenshot showing the performance of inertial and visual tracking by visualizing building wireframe models. At the bottom of the screenshot the created panorama is visualized.

6.10 Combination of attitude with visual tracking

This section presents results from applying the multi-sensor fusion system at the outdoor test site with disruptions from electromagnetic fields. In this scenario, the user went on-site to a location, held the AR platform in her hands and was performing orientation movements.

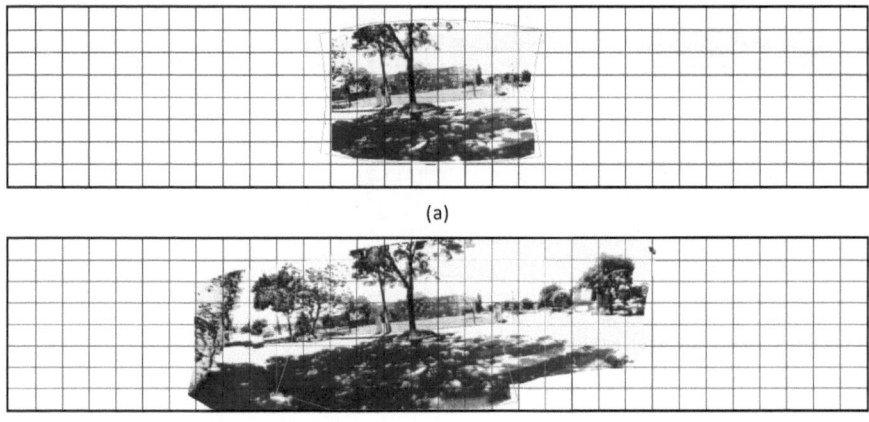

(a)

(b)

Figure 60: Map of the outdoor environment created by the visual panorama tracker. (a) First image used for mapping the test site and calculating the features. (b) After rotational movements the tracker has mapped a larger area of the environment.

The user's orientation was tracked using the state machine (described in Chapter 6.5), which combines the orientation estimates of both the Kalman filtered inertial sensor and the visual tracker. Nearly 180 degrees of the horizontal panorama was mapped, for the full length of the image in Figure 60 representing 360 degrees. Simultaneously, the Attitude Kalman filter delivers improved orientation estimates. Figure 58 shows results from the multi-sensor fusion approach, plotting the orientation values from both visual tracking and inertial tracking, together with the state of the fusion state machine. At the beginning, both trackers are valid and combined for calculating the final orientation. Within frames #25 to #75, the user rotates the handheld device. Both trackers continue to deliver accurate estimates. During frames #120 to about #210 the inertial tracker experiences transient deviation, caused by electro-magnetic influences. Now, the visual tracker provides the final orientation. After the deviation disappeared, both trackers are combined again. At sample #345 the user performs fast, abrupt rotations. Hence, the video camera delivers blurred images, which causes the visual tracker to fail. Now, the inertial orientation is taken as final orientation.

Table 7: Yaw measurements of visual tracker.

| Deg | Yaw of Visual tracker[deg] | | Measured Yaw of Visual tracker [deg] |
	Mean	Std dev	
-120	-118.680	0.11	0.0
-90	-89.200	0.07	29.48
0	0.002	0.01	89.20
90	88.960	0.32	88.94
120	118.050	0.17	29.09

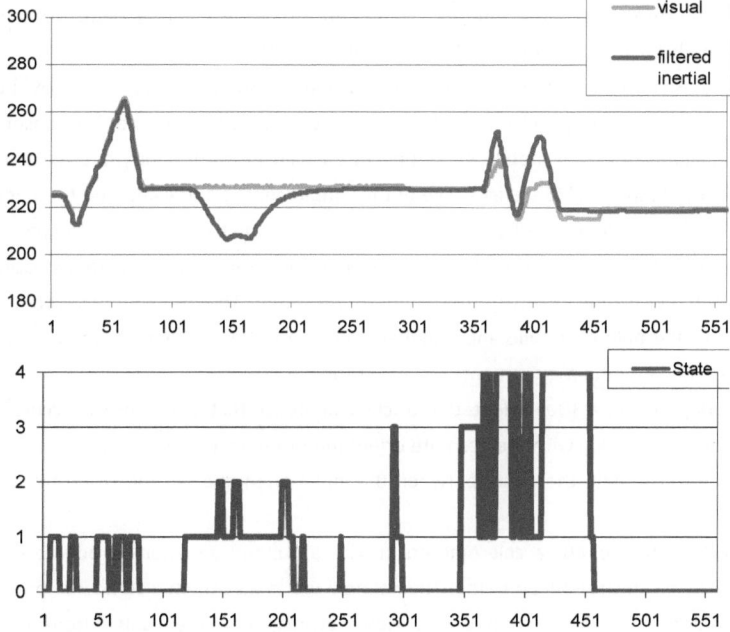

Figure 61: Comparison of yaw of the Kalman filtered inertial sensor with yaw of the visual panorama tracker under various conditions.

The user rotates the handheld device back again and near sample #460 the visual tracker re-initializes and continues tracking. Switching states depends on thresholds, but overall performs well.

The bars underneath the diagram in Figure 61 indicate which tracker is used for calculating the final orientation, which is used for the augmentation. The lower bars indicate that both trackers are used. In the upper diagram the lighter bar stands for visual tracking (V) and the darker bar stands for filtered inertial tracking (I).

6.11 North-centered orientation tracking

In the work presented in the previous sections, it was investigated how an orientation tracker can allow for correction of distortions of the magnetic compass and increase the robustness and accuracy of the pose estimates. However any initial distortion in the magnetic sensor would not be reduced over time. The approach described in this paper estimates the difference over time and can therefore reduce larger distortions in the compass and is based on (Schall, Mulloni, & Reitmayr, 2010).

With the rise of handheld AR systems such as Wikitude (Breuss-Schneeweis, 2009) built-in sensors are used to overlay registered information over a video background. The sensors mirror the modalities of earlier mobile AR setups, but usually perform poorly due to the use of cheap and low-power MEMS devices. The author implemented and evaluated the approach on a handheld device in terms of absolute orientation from the true north and compared it to a platform using high-end sensors. This comparison is specifically interesting for assessing the differences in registration quality of these types of AR setups. The results of this work directly contribute to the field of mobile and handheld augmented reality, since both accuracy and stability are fundamental requirements for registration in AR.

The work proposes a 3DoF orientation tracking approach that combines the accuracy and stability of vision tracking with the absolute orientation from inertial and magnetic sensors. A filter is used to estimate the offset between the initial orientation of the vision tracker and true north.

Magnetic compasses and accelerometers provide absolute estimations of orientation with respect to the earth's reference frame. Their simple use makes them a standard component in most outdoor augmented reality setups. However, magnetometers suffer from noise, jittering and temporal magnetic influences, often leading to deviations of 10s of degrees in the orientation measurement. Electrical installations, large objects made from conductive materials and even a user wearing a metal watch can bias the compass by several degrees. While dedicated off-the-shelf orientation sensors have improved steadily over the last decade, mobile phones must rely on less accurate components due to price and size limitations. Many outdoor AR applications now operating on mobile phones suffer from the inaccurate and jittering estimation of orientation with respect to the north. Visual tracking has become a corner stone of high quality AR systems providing pixel accurate overlays in video-see-through systems. Most solutions require a known model of the environment to provide camera orientation and position. For generic outdoor environments, such a model may not always exist.

Recent work in our group has focused on an efficient orientation tracking and mapping technique of a handheld device relative to an unknown starting orientation. As a result, annotations created and rendered relative to the mapped panorama are displayed accurately and steadily in unknown environments. Without knowledge of the global registration of a mapped panorama, the visualization of landmarks and information modeled in an earth reference frame is not possible. A straightforward use of the magnetic compass and linear accelerometer to register the mapped panorama leads to the aforementioned alignment errors and jitter. To overcome the limitations of these sensors, a 3DoF tracker was implemented that fuses sensors and vision-based orientation tracking. This approach is to treat the vision-based tracking as the main modality and to register the underlying panorama with respect to a north-down reference frame. To achieve this registration, the relative orientation between the vision-based tracker and the sensor-derived rotation over time is estimated.

Two implementations of this approach on two different types of hardware are presented. On a high-end configuration, a tablet PC and a sensor board was used. On a low-end configuration, a modern smart phone with an integrated compass and accelerometer was used. The author evaluated the accuracy of both systems in comparison with each other, and in comparison with the results that can be obtained without vision tracking. Accuracy results in terms of absolute orientation from the true north are shown. Moreover, accuracy measurements were performed while aiming the devices at different known reference points, representing the ground truth.

6.11.1 Orientation estimation and sensor fusion

The redundancy and stability provided by the visual tracker can help in coping with short-term influences thereby strongly reducing the influence of magnetic compasses to disturbances. The noise in the output of the smart phone compass (heading, s = 1.31deg) is one order of magnitude larger than the noise in the accelerometer output (pitch, s = 0.13deg, and roll, s = 0.23deg). The principle behind the combined orientation tracker is to continuously refine an online estimation of the relative orientation between the visual tracking component and the world reference frame. A north-oriented world reference frame N given locally by the direction to magnetic north and the gravity vector is assumed. The inertial and magnetic sensors measure the gravity and magnetic field vectors relative to a device reference frame D. The output of the sensors is then a rotation that maps the gravity vector and the direction of north from the world reference frame into the device reference frame.

The second tracking component is a visual orientation tracker that estimates a panoramic map of the environment on the fly. Like the sensors, it provides a rotation of the device R_{DP} from the reference frame P of the panorama into the device reference frame D.

In principle the device reference frame can differ for the camera and the sensors; however, assuming a calibrated device, two reference frames can be identified. The aim is to estimate the invariant rotation R_{PN} from the world reference frame N to the panorama reference frame P (see Figure 62 (left)). Composing the rotations from world to panorama to device reference frame the following is obtained

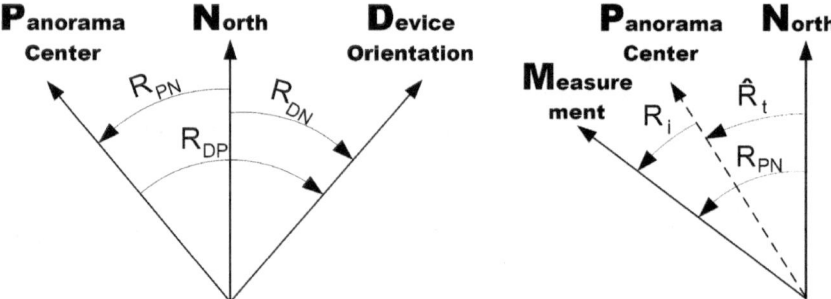

Figure 62: Reference systems. (Left) Overview of the rotations between world reference system N, device reference system D and panorama reference system P. (Right) Innovation motion given the Kalman filter's status and a new measurement.

$$R_{DN} = R_{DP} R_{PN} \Leftrightarrow \qquad (16)$$
$$R_{DP}^{-1} R_{DN} = R_{PN} \qquad (17)$$

Using equation (17) the relative rotation R_{PN} can be estimated online during operation of the system.

To estimate the orientation from sensor measurements, a simple procedure is followed. At a timestamp t, measurements g_t for the gravity vector and m_t for the magnetic field vector are received. A rotation $R_{DN} = [r_x r_y r_z]$ is calculated such that

$$g_t = R_{DN}g \text{ and,} \qquad\qquad\qquad (18)$$

$$m_t r_z = 0 . \qquad\qquad\qquad (19)$$

The resulting rotation accurately represents the pitch and roll measured through the linear accelerometers, while the magnetic field vector may vary within the plane of up and north direction (X-Y plane). This reflects the observation that the magnetic field vector is noisier and introduces errors into the roll and pitch of the device. The columns of R_{DN} are computed as

$$r_y = \frac{g_t}{\| g_t \|}, r_z = \frac{m_t}{\| m_t \|} \times r_y, r_x = r_y \times r_z . \qquad (20)$$

For the video frame available at the timestamp t the vision tracker provides a measurement of the rotation R_{DP}. Given the two measurements R_{DN} and R_{DP}, R_{PN} can be computed through the measurement equation (17).

6.11.2 Kalman filter setup

An extended Kalman filter (EKF) estimates the 3 parameters of the rotation R_{PN} using the exponential map of the Lie group SO(3) of rigid body rotations. The filter estimate at time t is represented as a rotation $\hat{R}_t$ that is related to the real R_{PN} through the following relation

$$R_{PN} = \exp(\mu)\, \hat{R}_t \text{ where } \mu \sim N(0, P_t). \qquad (21)$$

The covariance P_t describes the filters uncertainty about the state at time t. As a constant is estimated, a constant position update model is used with a small process noise sp to account for long-term changes in the environment. The prediction equations are then

$$\tilde{R}_{t+\delta} = \hat{R}_t \qquad \text{and} \qquad\qquad (22)$$

$$\tilde{P}_{t+\delta} = P_t + \sigma_p^t \delta I_3 . \qquad\qquad (23)$$

In the following the subscripts t are dropped for clarity. To update the filter with a new measurement R_{PN} (see Figure 62 (right)), a small innovation motion R_i is computed from the prior filter state rotation $\tilde{R}$ to the measurement rotation R_{PN} as

$$R_i = R_{PN} \, \tilde{R}^{-1} \, .$$
(24)

The logarithm $r_i = \log(R_i)$ of the innovation motion is effectively the innovation of the error m in the state representation. The derivative of the innovation with respect to the state is the identity I3 and the Kalman gain K is computed simply as

$$K = \tilde{P} \, (\tilde{P} + M)^{-1} \, ,$$
(25)

where M is the 3×3 measurement covariance matrix of R_{PN} transformed into the space of R_i. The posterior state estimate is then given by weighing the innovation motion with the Kalman filter gain K and multiplying it onto the prior estimate

$$\hat{R} = \exp(K \log(R_i)) \, \tilde{R} \, .$$
(26)

The posterior state covariance matrix P is updated using the normal Kalman filter equations. The global orientation of the device within the world reference frame is computed through concatenation of the estimated panorama reference frame orientation R_{PN} and the measured orientation from the visual tracker R_{DP} as described in equation (16). The accurate, but relative orientation from visual tracking is combined with a filtered estimate of the reference frame orientation.

6.11.3 Results

The approach was implemented on two different setups, representing high-end and low-end types of mobile augmented reality hardware. For the high-end setup, a tablet

PC was used; for the low-end setup, a common smart phone was used. Size, cost and weight of the two setups suit different types of end-users: a large but powerful setup for field professionals (e.g. maintenance personnel) and a pocket-size setup for occasional users. The vision-based orientation tracker used for the experiments is presented in detail in (Wagner, Mulloni, Langlotz, & Schmalstieg, 2010).

How did the experimental setup look like? The author tested the absolute accuracy of the hybrid orientation tracker using a set of surveyed reference points which are known for centimeter accuracy. Figure 63 (right) depicts the test site in a bird's eye view, highlighting the position of the reference points used as target points (1-8) and the reference point used for positioning the AR setups (RP).

A realistic outdoor test site is used in terms of distribution and distance of the aiming points serving as ground truth, since it represents a typical scenario for the AR applications (see Table 8). The author mounted the setups onto a tripod positioned above the reference point RP.

Table 8: Angle (in degrees) to magnetic and true north of reference points 1-8, as seen from the reference point RP.

Reference point	Angle to magnetic north	Angle to true north
1	322,76	325,74
2	352,08	355,06
3	5,26	8,24
4	67,25	70,23
5	91,32	94,30
6	112,34	115,32
7	244,12	247,10
8	267,79	270,77

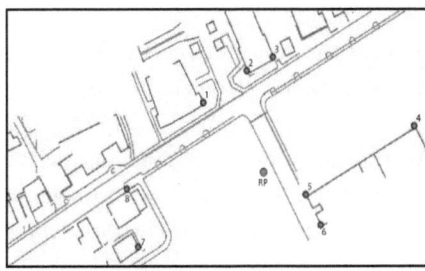 

Figure 63: Test area. (Left) north-aligned map of the test area and (right) bird's eye view of the test area, showing the position of all reference points.

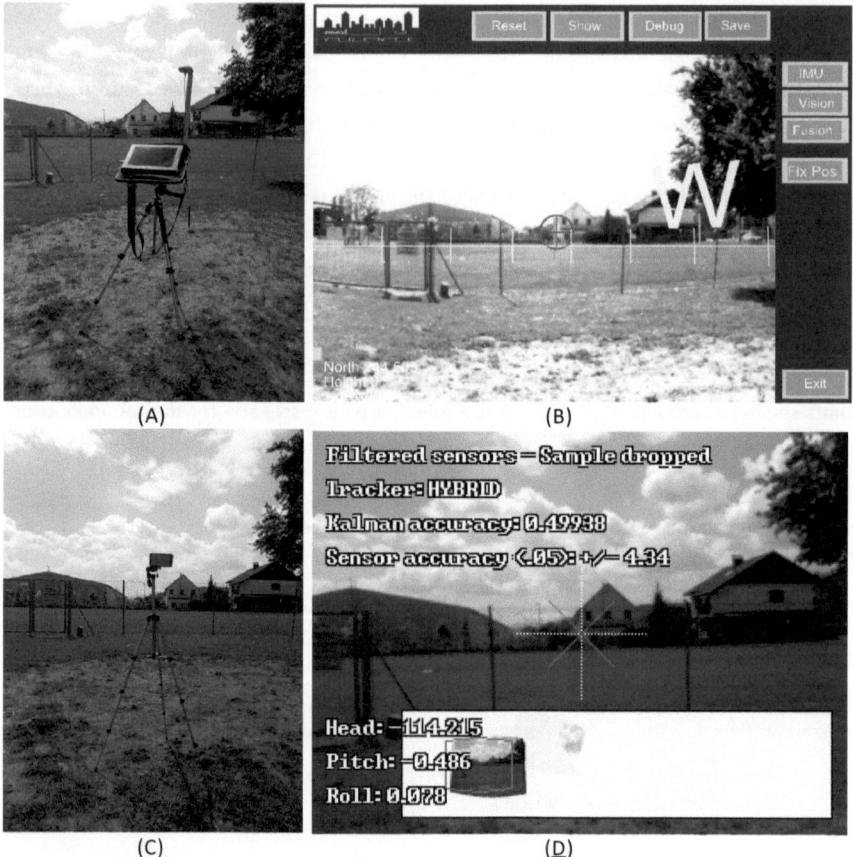

(A)   (B)

(C)   (D)

Figure 64: Test setup. (A) Physical mounting and (B) screenshot of the tablet PC-based and (C-D) of the phone-based setups used for accuracy measurements.

The accuracy of the tracker was measured by aiming the device's camera at one of the reference points and subsequently turned the device towards all other reference points without resetting the tracker.

The device was kept still for about 30 seconds at each reference point, logging the orientations reported by the sensors, by the vision tracker and by the hybrid tracker. A viewfinder glyph was visualized on the device's screen to ensure pixel-accurate alignment of the camera with the real-world reference points in the environment (see Figure 64).

Next, take a look at the accuracy results of the experiments. Figure 67 depicts a plot of a measurement session turning the tablet PC in clockwise direction, from one reference point to the next. Since the visual tracker does not provide absolute orientation from the north, in the plot it is assumed that it has zero error on the first sample. The results demonstrate two improvements over a pure sensor-based orientation tracking. Firstly, high frequency noise is reduced. The visual tracking is dominating the motion estimation and provides a low jitter rotation estimate. Secondly, over time, the error of the filtered rotation is smaller than the sensor-only rotation because deviations in the compass measurements are averaged over different orientations. Overall, a responsive, less jittery estimate is obtained that, on average, is also more accurate than the orientation derived from the sensors alone. Figure 68 shows the plot of a similar measurement session using the mobile phone, while Figure 65 (left) shows the error to the closest reference point. For the mobile phone setup, similar results are obtained as for the tablet PC. A notable difference is that the measurements here start with a bias of several degrees. This initial bias can only be removed after other directions have been visited and towards the second half of the session, the hybrid tracker is able to correct the deviation.

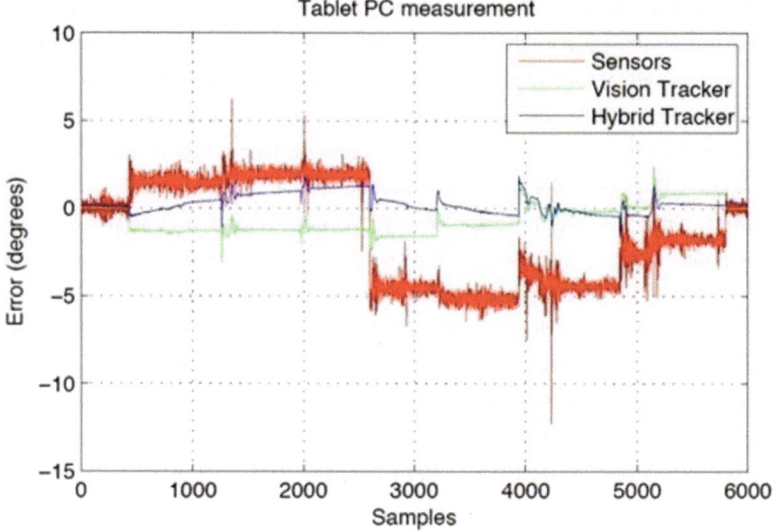

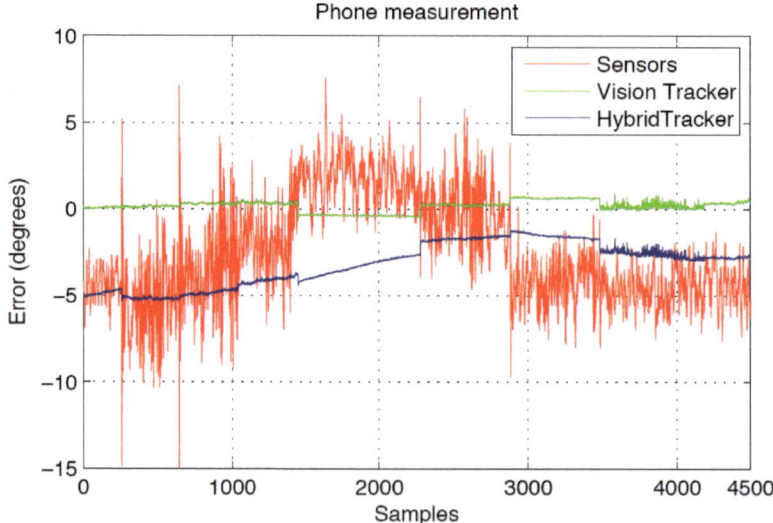

Figure 65: Errors (in degrees) to the north for the sequences recorded on the tablet PC (top) and on the phone (bottom).

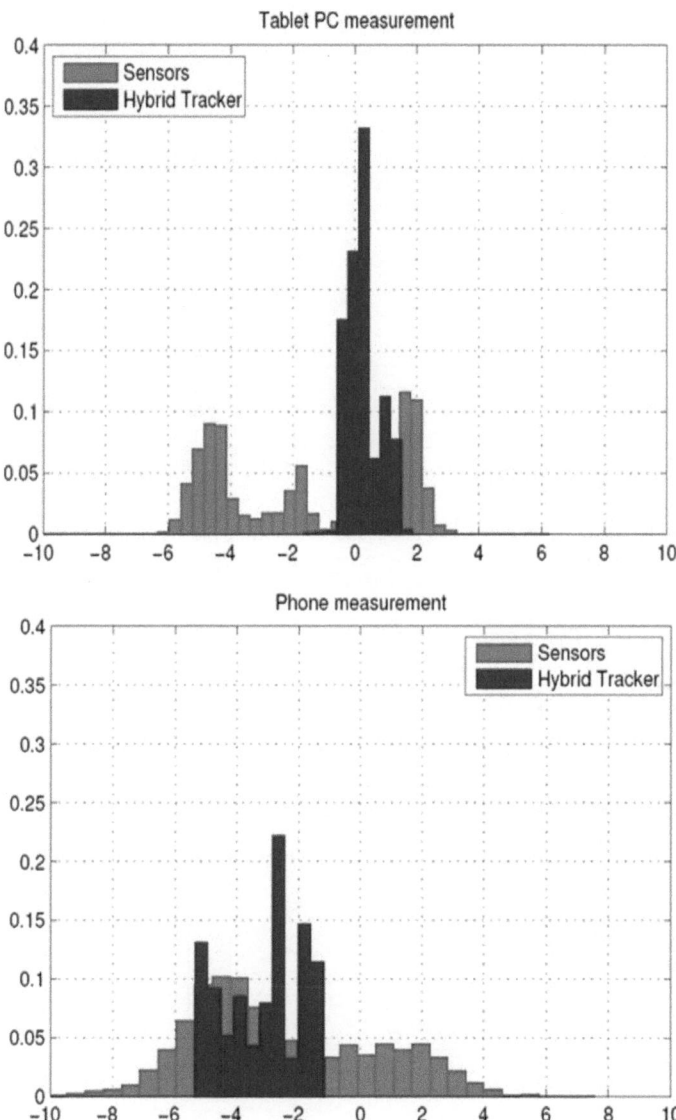

Figure 66: Error distribution from the tablet PC (top) and for the phone (bottom), for both the sensors and the hybrid tracker.

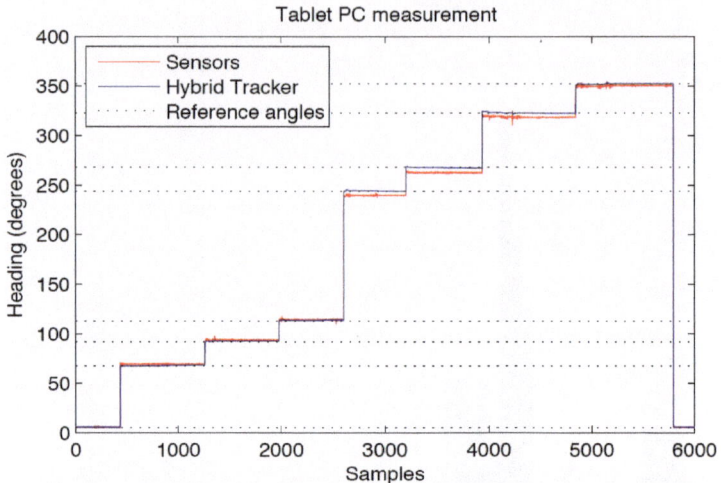

Figure 67: Test sequence for the tablet PC. Recorded test sequences showing the headings for sensors only and hybrid tracker.

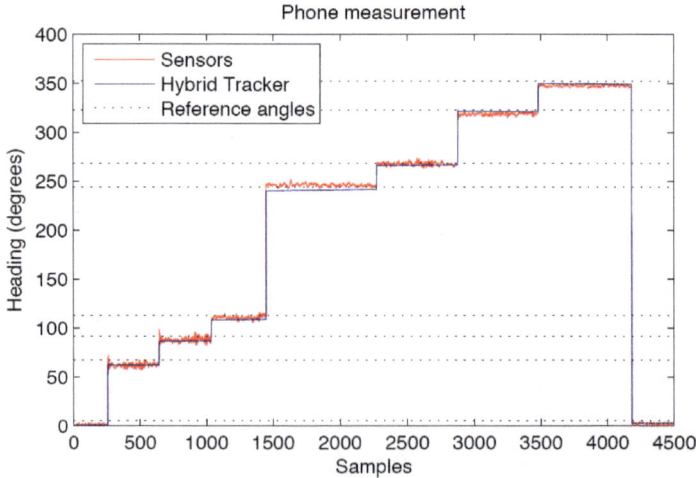

Figure 68: Test sequence for the phone. Recorded test sequences showing the headings for sensors only and hybrid tracker.

Table 9: Mean and standard deviation (in degrees) for the error of the sensors, the vision tracker and the hybrid tracker from both the tablet PC and the phone measurements.

	Tablet PC		Phone	
	Mean	Std dev	Mean	Std dev
Sensors	-1,35	2,80	-2,53	3,00
Vision tracker	-0,62	0,80	0,19	0,37
Hybrid tracker	0,26	0,50	-3,15	1,25

The hybrid tracker effectively averages over all these errors and produces a better mean estimate. Figure 66 shows the error distributions from the tablet PC and the phone for both sensors and the hybrid tracker. One can observe that the distribution for the tablet PC is broader and has a higher peak.

While the previous experiments were performed with the device mounted on a tripod, the following results show the behavior when using the devices in a free-hand manner. While the tripod-based accuracy evaluation shows a measure of absolute accuracy of our hybrid tracker, free-hand motion (holding the device in the hand) shows the tracker's behavior in a more realistic and dynamic scenario. Figure 69 shows a plot of rotating the tablet PC from one reference point to another (represented by the two dotted lines), through a natural rotational movement. The raw sensors are compared with a Kalman filter running on the sensors input and tuned for low latency and reasonable filtering of high-frequency noise. Also the output of the hybrid vision sensors tracker is presented. The plot shows clearly that the hybrid tracker does not suffer from the latency of the filtered estimates. Also, thanks to the redundancy given by the vision tracker, the hybrid tracker is able to eliminate the false changes in pitch and roll due to the accelerometer measuring both gravity and the centripetal acceleration of the device.

The proposed tracking approach increases robustness through the redundancy given by visual and inertial orientation estimates. The results of this work directly contribute to the field of mobile and handheld augmented reality, since both accuracy and stability are fundamental requirements for registration in AR.

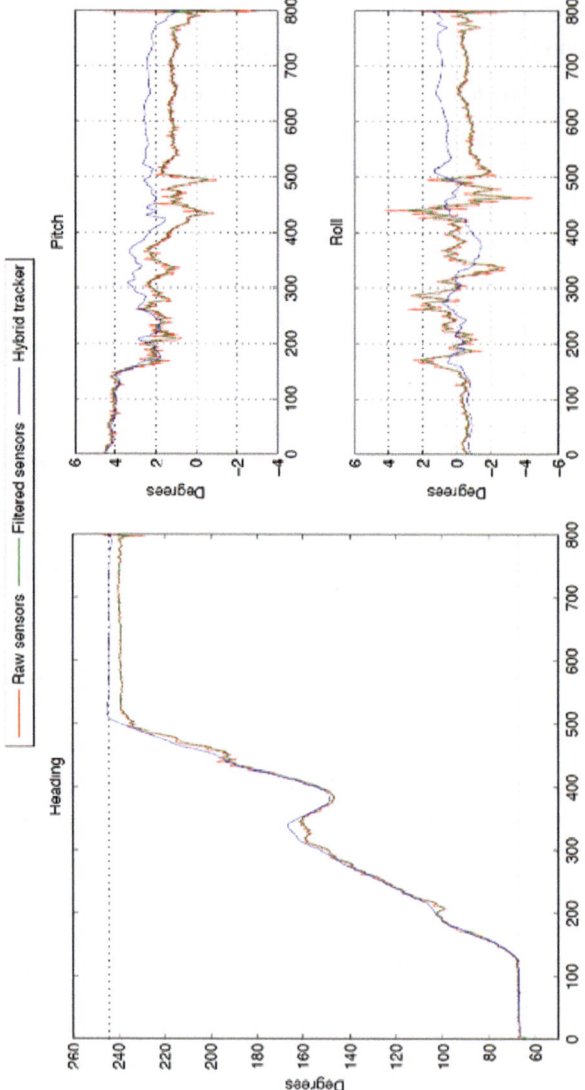

Figure 69: Plot of heading, pitch and roll for a free-hand movement of the tablet PC between two reference points. Orientation for the raw sensor values, a filtered estimate and the hybrid tracker are plotted.

6.12 Discussion

Generally, mobile AR applications require accurate and global pose estimation. Several techniques and methods that contribute to improved accuracy, stability and robustness for pose estimation were presented.

A convincing tracking solution must overcome the inherent limitations of individual tracking techniques by combining different complementary methods. Consequently, multi-sensor fusion approaches were implemented. In addition to the 3DoF positioning systems such as RTK GPS in outdoor environments (or Ubisense UWB indoors) and the video camera, an IMU is used as an additional sensor. While the orientation information from the IMU and the position information from the 3DoF positioning system only provide complementary measurements of the camera pose, the video stream encodes relative motion information about both translation and rotation. During operation, the proposed system records IMU measurements, GPS measurements and video frames taken at the same time. Motion estimation and feature matching between pairs of video frames create epipolar constraints on the camera motion between the frames.

Inertial tracking has the advantages of range and yields a system that is passive and self-contained. Its major disadvantage is its lack of accuracy and drift over time. The first effect of time-dependent drift of the accelerometers angular rates can be corrected by an Attitude Kalman filter which performs a sensor fusion of gyroscopes, accelerometers and magnetometer. Moreover, the gyroscopic angular rates have biases which are estimated in this filter as well in order to avoid a temporal drift of the attitude angles. This filter is also capable of eliminating the rather long transient oscillation behavior of the inertial sensor. The second effect of location-depended deviations of yaw can be detected and corrected for by using a visual tracker which does not require a model of the environment. Through online mapping and learning of natural features of the unknown environment, this tracker allows for detecting and correcting the deviation of the 3-axis compass. This improves both the accuracy and the robustness of the orientation estimates. This makes the rotation much more stable with respect to the real world than normal inertial tracking which typically has some lag, drift and/or slight misalignment.

Experiments were conducted to asses the accuracies of the single sensors as well as the performance of the multi-sensor fusion tracking approach. While providing sub-meter accuracy position estimates using the coarse 3DoF tracking system, the accuracy and robustness of the orientation estimates of the mobile device could be increased significantly under real-world conditions.

Die lohnendsten Forschungen sind diejenigen, welche, indem sie den Denker erfreun, zugleich der Menschheit nützen.

Christian Doppler, 1803-1853

7. Applications – AR visualization and interaction in civil engineering

The following section focuses on outdoor AR. Using the hardware prototypes presented in Chapter 5 a variety of research experiments and evaluations were performed in outdoor environments. The prototypes were the basis of experimentation with visualization of geospatial data and interaction with the data in the field with the use of GIS data. The data for the experiments were generated using the transcoding pipeline presented in Chapter 4.4.

An outdoor AR system relying on GIS data can be considered to be a special case GIS. It presents geo-referenced information in real-time and in 3D based on the physical location of the user, user preferences and other context-dependent information. Large amounts of geo-referenced information, for example a 3D world model, require a database system for efficient storage and retrieval. The introduction of a GIS database also solves the problem of providing a consistent view of the 3D world model for a potentially large number of wirelessly connected clients.

For improved efficiency, paper maps are increasingly being replaced by notebook computers taken to the field to directly consult the GIS. A GIS database normally employs two-dimensional models to represent the geographic data. Accurate evaluation of a situation from a map and a GPS location requires applying a mental transformation from map to reality. This assumes that the user is familiar with the significance of map scale, generalization and symbol language. In many cases this cannot be taken for granted. Even users experienced in map-reading may struggle if for example reference surface features are occluded by winter snow. AR thus has the potential to remove the need for a mental transformation from map to reality.

Workers in the field have a strong need to locate their assets, for example structures scheduled for maintenance or to ensure safety for digging at excavation sites. Among the procedures that can benefit from employing AR in field work are contractor assistance, outage management and network planning. Simple localization is important for the on-site informing of contracting staff. For this aim, a registered AR view can provide fast and accurate localization of subsurface assets, thereby reducing risks of accidentally damaging underground infrastructure during excavation. An important task in outage management is the detection of gas leaks and cable damage. Workers must trace a trench with special sensors such as a "gas sniffer". Navigation along the trench with a mobile GIS is rather cumbersome. AR can provide a superior graphical overlay view, outlining the trench to follow and highlighting relevant underground assets.

Planning of utility networks is usually done in a planning office using desktop GIS. A plan for a new trench has to be verified on location before being submitted to the responsible authorities. This task is traditionally accomplished by taking paper maps to the field and annotating them. AR means that planners can be provided with a graphical overlay of the planned trench and can directly modify the plan to incorporate required changes using mobile spatial interaction tools without the need for any post-processing. The trend in the geospatial community clearly points in the direction of mobile GIS.

7.1   Concept

The purpose of maps, geographers know, is to model reality. In the *Nature of Maps* (Robinson & Petchenik, 1976) defined a map as a "graphic representation of the milieu". The use of the term *milieu* is interesting because it suggests much more than the flat, static maps users are familiar with. It presents a challenge to step beyond the comfortable reach of 2D representations to higher dimensions of visualization. To model reality most clearly, it certainly makes sense that users strive to map what they actually experience. Today, the established way to use GIS in the field is through paper plans, which are plotted as needed and manually annotated on a construction or maintenance site if changes are made. There is a certain trend towards 3D GIS that has not evolved as much as the area of 3D visualization. However in the utility sector, the need to work with paper plans and the fact that underground assets are normally hidden has limited the interest in 3D GIS. Nevertheless, the real environment visited by field workers is still three-dimensional.

Figure 70 shows a mobile user augmenting the real environment with geospatial 3D model. This experiment was done with the first working prototype for that purpose in 2006. However registration is very poor due to inaccurate GPS tracking.

Figure 71 shows a mobile user with more sophisticated mobile AR setups. Furthermore, Figure 72 gives an example of how such a 3D scene overlaid on the real-world construction site looks like. The application clearly aims at close-up investigations of assets and at inspections in the immediate environs of the field worker's current position.

Figure 70: AR visualization of GIS data. (Left) User with mobile AR system. (Right) Users view of very simple visualization of geospatial data.

Figure 71: User with mobile outdoor prototypes. (Left) Vesp´R based setup. (Right) Tablet PC-based setup.

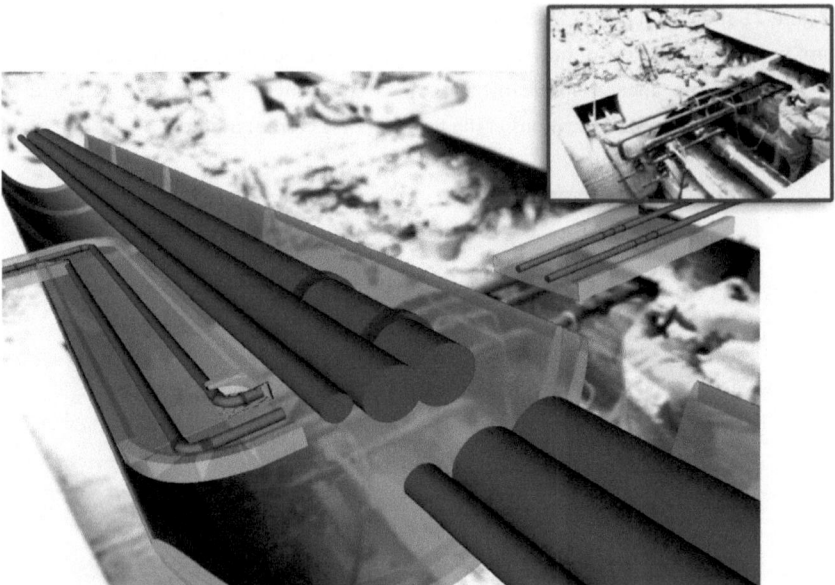

Figure 72: Screenshot of a working outdoor prototype of the project Vidente. Data was extracted from a geodatabase and automatically transcoded onto a scene-graph data structure.

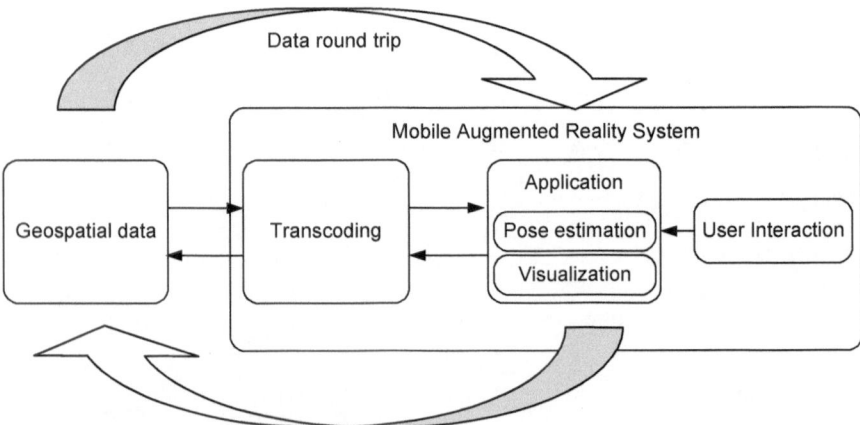

Figure 73: Smart Vidente system architecture.

When starting with the implementation of the mobile AR framework, solutions for 3D GIS visualization were not sufficiently advanced for the authors purposes. Consequently, an own XML based data format and processing pipeline using XSLT for data translation was developed. Recently, the emergent GML3 standard together with the WFS standard provide a standardized and extensible interface for accessing GIS information. The three-dimensional geometry must be extracted from a conventional database system and interpreted on the fly as a 3D visualization. However the three-dimensional models are not stored persistently. Rather, the underlying data is stored persistently and managed in the utilities' geospatial databases. Hence, always the most current data version is accessed and users can benefit from all the advantages of a powerful database system such as data versioning, loss prevention, recovery, integrity enforcement and comfortable operations for retrieval, insertion and update. Data redundancy and inconsistency among spatially overlapping models are eliminated since all models refer to a common data source. A lean and generic GML application schema (VidenteGML) serves to encode the underlying geo-referenced utility asset data issued from the data server as shown by (Junghanns, Schall, & Schmalstieg, 2008). Vidente (Schall, Junghanns, & Schmalstieg, 2010) is based on a multi-tier system architecture with a mobile front-end and an operational geospatial database as a back-end (Figure 73). The mobile front-end is a handheld client device, which is designed as video see-through. Hence, scenes are assembled at the client device in real-time by merging continuously streamed video footage with geo-referenced computer graphics considering the client's currently tracked position and orientation. Registration in 3D requires the capability to perform accurate global localization and pose tracking in real-time. A handheld setup was equipped with tracking sensors designed for outdoor use.

Among others these tools comprise data retrieval capabilities, redlining functionality to annotate the geospatial assets and a virtual excavation tool to improve depth perception of complex underground infrastructure. In particular useful is the planning tool, which allows for visualizing projected assets superimposed over the real world. Moreover, the position of the projected asset can be changed on-site interactively.

7.1.1 Inspection toolset

To optimize the benefit of an AR application, all presented information has to be de-signed towards an intuitively understandable visualization. However the simultaneous representation of both virtual and real information introduces a number of difficulties. For example, virtual data always overrides real-world information, which is in particular problematic when presenting subsurface structures in so called X-ray visualizations.

Figure 74: Image-based ghostings. (Left) Careless overlay with virtual content, where occlusion cues are not available. Middle: Ghosting map created by this technique. (Right) X-ray view of the same scene using image-based ghostings. The image-based ghosting preserves essential perceptual cues that help to understand the relationship of depth between hidden information and the physical scene.

Therefore, virtual and real information have to be carefully chosen to avoid problems of depth perception, caused by a loss of information. To handle this problem, work has been shown focusing on the modifications of hidden structure, while other research was concentrating on the stylization of the occluding objects (e.g. pavement boarders occluding subsurface electricity lines) (see Figure 75). For example, (Bane & Höllerer, 2004) present interactive tools to select certain parts of hidden geometry. In the following, a series of tools are presented that were implemented for visualization and interaction purposes with underground infrastructure models. Moreover, in very recent work (Zollmann, Kalkofen, Mendez, & Reitmayr, 2010) show an image-based ghosting approach for visualizing hidden structures (see Figure 74).

7.1.2 Excavation tool

Indiscriminately overlaying hidden information on top of visible real-world entities introduces depth perception problems. Virtual objects appear to float on top of the real ones because of overdraw. Therefore an excavation tool resembling a hole in the ground is used, thereby providing plausible interpretation of depth through partial object occlusion as well as motion parallax (see Figure 76). The excavation tool is implemented using a magic lens technique, filtering the content based on contextual information derived from the attribute data in the GIS (Kalkofen, Mendez, & Schmalstieg, 2007). The lens is initially positioned in front of the user, but can be adjusted using controls on the AR device.

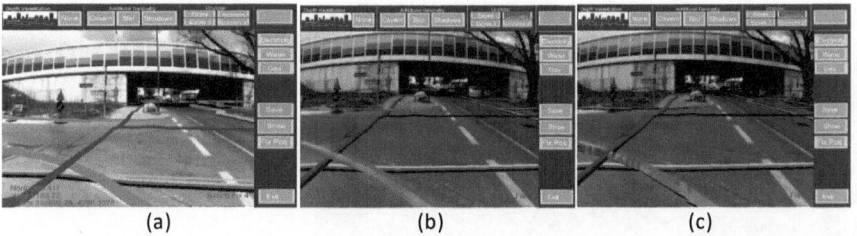

(a)   (b)   (c)

Figure 75: AR view showing underground pipes well registered in 3D. Pavement boarders are used as occluders for the pipes. (a) Simple visualization. (b) Pavement rendered transparently. (c) Pavement rendered using perlin noise.

Figure 76: Excavation tool (left). Screenshots demonstrating improved depth perception by adding specific depth cues. (Right) Metadata querying tool. Using a cross-hair a user can select the water line and query related semantic information (screenshot) (Data courtesy of Graz AG - Stadtwerke fuer kommunale Dienste)

7.1.3 Metadata Querying tool

Also a metadata querying tool was implemented, which helps the user to visualize the meta information of the infrastructure, such as part number, ownership etc. This meta information is obtained from the original geo-data and stored as non-geometrical attributes on the 3D model. As depicted in Figure 76 (left) a crosshair target can be positioned on top of an asset, revealing associated meta-information.

Figure 77: Snapshot of the augmented live video. Augmented
snapshot is stored for documentation purposes.

7.1.4 Filtering tool

Desktop GIS systems offer advanced possibilities for filtering and selecting information
to avoid cluttering. Such detailed attribute selection tends to be too complicated for
interaction in handheld AR. Instead, the user can select a region of interest with the
excavation tool first, and then turn on 3D features based on pre-grouping into asset
categories (gas, water, buildings and so on). This two-step filtering approach reduces
clutter to a manageable amount with only a minimum of interaction.

7.2   Snapshot tool

For documentation, field workers like to freeze an image at any point in time and take a
snapshot, to be analyzed later in the planning office. A dedicated button on the AR device triggers such a snapshot (see Figure 77).

7.3   Interactive redlining toolset

The established way to deliver geospatial data outdoors is through plotted paper plans.
The plans are manually annotated directly on the construction or maintenance site, if
changes are made — this procedure is often called redlining.

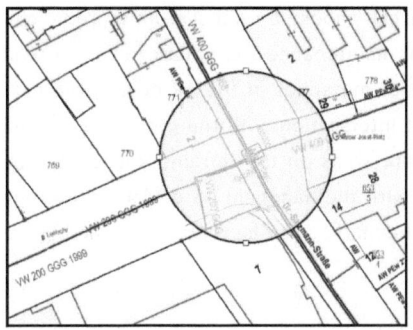 

Figure 78: Conventional redlining feature in 2D (left). Visualized in a conventional geographic information system (GE Smallworld™). (Right) AR redlining feature shown in 3D. It is directly superimposed on the street level using the augmented reality visualization.

For improved efficiency, paper plans are increasingly replaced by notebook computers taken to the field to directly consult the GIS. Just like the paper plan, the GIS uses two-dimensional models to represent the geospatial data.

However geospatial objects in the GIS can directly be annotated with virtual redlining which is a very important operation that bridges the gap between the office and the field, for example in planning and in network inspection. Planning of utility networks is usually done in an office using a desktop GIS.

A plan of, say, a new trench has to be verified on site before being submitted to the corresponding authorities. This task is traditionally accomplished by taking paper maps to the field and annotating them. Utility companies also need to inspect their network on a regular basis to evaluate its condition. During the network inspection, every subsurface asset is rated and notes are taken by the field workers. Using a notebook computer running the GIS in the field allows entering redlining information directly. However accurate judgment of a situation from a map requires applying a mental transformation from map to reality.

This assumes that the user is familiar with the significance of map scale, generalization and symbol language. In many cases this cannot be taken for granted. Even users experienced in map-reading may struggle if, for example, reference surface features are occluded, such as with winter snow.

7.3.1 Annotating the geospatial model

A redlining tool in an augmented reality style was implemented for providing field workers redlining capabilities. The redlining tool enables the outdoor user to annotate and interact with geospatial objects. The user can choose a symbol from a predefined palette of symbols (e.g. damage, safety area, or maintenance area) that can be placed in the geospatial model. Using the point-and-shoot metaphor of the AR device, the user can place the selected symbol at the point of intersection with the geospatial model. Furthermore, the tool enables the user to mark areas on the terrain. This is done by choosing the centre of an area to be marked with the handheld AR device again using the same metaphor. The radius of the area to be marked can be changed by varying the pitch and yaw of the device (see Figure 78).

7.3.2 Surveying in the geospatial model

Besides placing an annotation to a location in the geospatial 3D model the user also has the possibility to survey locations by intersecting his pose with the geospatial model (underground infrastructure, DTM or buildings). This enables a field worker for example to survey a single spot or the location of a trench. Furthermore the user can draw a freehand polygon on the DTM and store its location. Small cubes indicate the surveyed locations by the field worker. Redlining information can be stored in a separate file (e.g. each location and symbol), for later analysis in the office.

7.3.3 Interactive validation of object placement

Some applications demand to inspect, validate or modify the placement of specific structures in the environment. This can be necessary if either the GIS is known to be incomplete, so that planning exclusively in the office is not feasible, or if plans from contractors are obtained without geo-referencing. In this case, surveying in the field and planning the actual location of the asset can be integrated in one interactive feature of the AR system, assuming that a 3D representation of the inspected structure is already available. For example, Figure 79 shows a noise protection barrier to be erected alongside a railroad track. The barrier has been planned by a contractor, while the exact placement of the barrier is subject to the on-site inspection with the AR system. The barrier must not be built on top of existing underground utility infrastructure, to assure that maintenance of the utilities is not affected. In order to do that, the field workers determine various possible placements in an on-site planning discussion.

Figure 79: Noise protection barrier to be erected alongside a railroad track. The planned barrier is subject to the on-site inspection with the AR system to determine overlapping areas with existing underground infrastructure.

7.4   Verification toolset

Next, a toolset is implemented that supports the user verifying geospatial data on-site.

7.4.1 Visualization of abstract information

Field workers from utility companies that are only using land-register data and no topographic data in their GIS have the strong need to localize the geospatial objects directly on-site. Typically this is done by using a real-time kinematics GPS device to obtain references to the real topography. Undoubtedly, this is a time consuming task. Visualization of legally binding land-register data is an important task for utility suppliers, since this information is usually difficult to find on-site. A wide range of abstract information, such as parcel borders, parcel areas, ownership and servitude rights (see Figure 80) are relevant for this task. Superimposing that data with the AR device does not need any further measurement steps, since the user knows his pose in relation to the real world topography due to tracking anyway. In particular utility companies that only employ land-register data need to determine and find the land-register data in the real environment. To do so, they have the need to transfer the parcel border or trench border from the land-register data to the real on-site environment. Consequently they survey the borders, because borders in the real world are not necessarily on the right place.

Figure 80: Conventional exocentric view at land register and underground services data as available in two-dimensional GIS visualizations (Graz Geodatenserver) (left). (Right) Egocentric view at land register (dark grey on the right highlighting extents of adjacent parcel) and underground services data using AR techniques.

For example, it cannot be assumed that a fence is exactly located at the parcel border. The mobile AR device provides the possibility to fulfill the positioning requirements by having an integrated GPS receiver. Figure 80 shows darker marked areas indicating land-register data superimposed on the real environment. In contrast, compare this 3D visualization with the one in 2D usually done on office PCs. Using the augmented reality style visualization the user sees the parcels in an egocentric view directly superimposed on the environment.

7.4.2 Verification of abstract information

Often field workers from utility companies need to verify data from the GIS on-site at construction or surveying sites. For example, in an excavation or digging task it is essential to not dig on wrong parcels. The verification toolset can support the field workers in the verification of land-register and cadastral data.

7.5   Evaluation results

In order to assess the applicability of the toolsets, a series of field trials and interviews were performed, analyzing a range of aspects. These factors included the general quality of the method, the matching of industrial requirements obtained in the system requirement phase, and the actual operation by end-users. The main focus was on the practical relevance of the prototype. By conducting an interview with field workers from industrial utility companies (two employees from the local power supplier E-Werk Gösting Stromversorgungs-GmbH and five employees from Salzburg AG operating gas, water, electricity and heating networks), valuable feedback from experts with several years of experience was obtained.

The question arose of how to evaluate a mobile AR application in outdoor environments in a meaningful way. Arguments were worked out considering the framework of (Olsen Jr, 2007) who investigated how to evaluate user interface systems that are off-the-desktop and nomadic. This will involve new devices and new software systems for creating interactive applications, such as LBS or mobile AR systems. Simple usability testing is not adequate for evaluating such complex systems. Olsen suggests a set of criteria or claims for evaluating new user interface systems. Every new piece of interactive technology addresses a particular set of users, performing some set of tasks, in some set of situations. It is critical that interactive innovation be clearly set in a context of situations, tasks and users. The STU (Situations, Tasks, Users) context forms a framework for evaluating the quality of a system innovation. Before all other claims a system or interactive technique must demonstrate importance. Importance analysis proceeds directly form the intended STU context. The first question is the importance of the user population (U). In this case the user population consists of field workers who are definitely important for a functioning infrastructure, which is a backbone of our modern lives. Next, the importance of the performed task (T) for the user population must be evaluated. Importance might be established by how frequently the task occurs. It might also be established by looking at the consequences of not being able to do the task. In this case field workers are performing inspection tasks, correction tasks and redlining tasks for infrastructure networks very regular on a daily basis. Undoubtedly, the consequences of not performing these tasks on infrastructure networks can be catastrophic. Furthermore, the importance of the situations (S) needs to be evaluated. How often do the target users find themselves in these situations and do they need to perform these tasks (T) in those situations? In this case the situations include maintenance and planning activities on the underground infrastructure network. Field workers regularly need to deal with situations such as difficult environments that assets are not obvious or that there is danger ahead. Also this third criterion for importance of a system interface can be answered with yes as well as the importance of the STU context as a whole.

According to Olsen, tools for creating new user interfaces can be improved by increasing the expressive match of the system. Expressive match is an estimate of how close the means for expressing design choices are to the problem being solved. There are several requirements when making a claim of grater expressive match. One must demonstrate that the new form of expression is actually a better match. Frequently greater expressive match is tied to a claim to lower skill barriers. In the case of the mobile AR system, Expressive Match is significantly increased, as AR provides a closer match to the real on-site situation by visualizing registered 3D models on-site. Moreover, the integrated 3D visualization provides a much more intuitive interface enabling users that are not familiar with reading maps, to assess the geospatial objects in their surrounding.

The objective "Simplifying Interconnection" is given if the new system under test can be better embedded in workflow than existing systems. This is true for the proposed AR system, because the workflow of typical tasks performed from field workers from the utility sector, can be significantly improved. In this context, also the possibility of a loss-less data round trip must be mentioned. Current tasks often involve printing digital information on paper maps, making annotations per hand, and typing these annotations into the GIS system when back in the office. The system under evaluation can provide a data round trip without a digital gap and consequently task can be performed more efficient.

7.5.1 Evaluation procedure

A semi-structured interview was performed evaluating the practical applicability of the tools and the usefulness of terrain models for their tasks. An early trial and interview was performed with five experienced field workers. All field workers confirmed the high potential of AR for time savings and error avoidance in tasks like construction instruction, outage management and planning. Most importantly the visualization overlaying the underground infrastructure over the real world needs to be highly accurate. High priority was given to depth perception of the buried assets which reconfirmed the expectations. Field workers expressed their wish to see all underground assets buried at one spot, allowing achieving a complete overview. It became evident that color coding for different bands — voltage bands for electricity or pressure bands for gas pipes — is highly desired, since it helps a lot in classifying the assets. Vidente can support that by choosing the color code according to the attribute values of the underground infrastructure. Furthermore, field workers mentioned photorealism of all rendered graphics not to be of primary importance. Concerning the user interface, also touch screen based interaction would be conceivable, since many people are used to control applications that way.

Communication between utility companies and construction companies is conventionally done by spraying markers on the ground. This can be seen as another form of redlining. Using the application, spraying is no longer necessary. The snapshot tool is very relevant, when a certain situation concerning underground infrastructure needs to be discussed, documented or presented.

E-Werk Gösting usually needs to locate 50-100 meters of trench length a day. The AR device would be operated in a discontinuously mode, using it for ~5-10 seconds, and then walking further. The overall time of usage at one construction site is around 15-20 minutes. The AR system can alleviate their work by carrying less measurement devices with them. It can even be used to digitize of a trench by walking along its course and recording GPS positions. The required positioning accuracy must be better than 30 centimeters in 3D. The location of interest, which needs to be determined with sub-meter accuracy (not much kinematic movement happens when the user arrives at the location). From this static location, the field worker inspects the underground infrastructure around him by scanning his surroundings with the handheld AR device, resulting in a circular motion. Overall, a smaller workload and fewer erroneous excavations are expected.

7.5.2 Digital terrain model

Including a DTM into the geospatial model was rated very helpful for visualization. As expected, the visualized terrain in particular was found useful in uneven areas, since the pipes need to follow the terrain. Without a DTM the registration in uneven terrain would only be correct in the near view (looking down onto the street). Additionally the DTM improves the reference of both underground features and above surface features like buildings.

Furthermore, two field workers from Salzburg AG mentioned when they only see the underground infrastructure without other assets superimposed on the real world they feel unsure if the registration is correct. Both DTM and buildings help people to rate the registration as reliable, because they first search for building lines in the model that are correlating to real building lines. After having found such references the location of underground infrastructure is trusted. This is consistent with the findings of (Robertson, MacIntyre, & Walker, 2008).

7.5.3 Virtual redlining

E-Werk Gösting currently works with analog plans in the field and work with both land-registers and topography data in the GIS. Their field workers need such redlining functions every day whereas the GIS is updated with redlining information on a weekly basis.

Currently the field workers do all redlining on plotted paper plans. In contrast, employees of Salzburg AG use tablet PCs in many processes, where redlining is a widely used technique. They currently use a mobile application from SAP for asset management and redlining purposes.

Business processes of utility companies. Field workers from Salzburg AG saw a practical application of the redlining tool in the network inspection process. The utility network is inspected on a regular basis, where a field worker needs to describe the condition of the single assets. Usually each asset is rated using a scale from 1 to 5. In this task redlining allows the field worker to assign the appropriate redlining symbols (1-5) to the assets. Furthermore, the redlining symbols with the according assets are then stored to a file and can be used for further tasks.

A second application area that was identified by the field workers is achieving of data. Utility suppliers have the need to archive all relevant information concerning each geospatial object. This is a time consuming task since the database must be kept up-to-date manually in the office. With the virtual redlining tool, it is easy to take snapshots of redlined geospatial objects, store and connect them to the according object. Salzburg AG stated that such a procedure would save a massive amount of work.

Experts from both companies mentioned the task of planning new geospatial objects or trenches as very useful for the redlining tool. As shown in Figure 78, a field worker from Salzburg AG had the task of surveying the locations of a newly planned pipe (in red color). The small cubes in the image represent the locations the field worker has surveyed using the AR device with the redlining tool. Additionally, the coordinates of these surveyed locations could be used for further processing.

Field workers from Salzburg AG have a demand in localizing street features, such as manhole covers or water openings. They often need to find the features again even if they have been covered (e.g. by a new thin layer of asphalt). By surveying such features on the street using the redlining tool takes advantage of the point-and-shoot metaphor they can easier find the features again.

Rating of redlining modes. The field workers were asked to rate the various modes of the redlining tool. Drawing of circular or rectangular areas with variable radius or size, as depicted in Figure 78 (right), was rated mediocre, despite the fact that it directly mimics a function available in the GIS. Interestingly, drawing rectangular areas received better feedback than circular areas, since many geospatial objects have rectangular shape (e.g. a trench).

Using a palette of predefined symbols was found very useful by all participants. This way they only need to place a specific symbol to a location instead of writing a description every time. For example, this mode simplifies the network inspection task in which the conditions of geospatial assets are determined and described with a number from 1 to 5. In this case a palette of symbols is sufficient allowing the user to place a symbol onto a geospatial object using the point-and-shoot metaphor.

For the task of describing and annotating assets (e.g. a damaged pipe) a palette of symbols gives too little information. In this case textboxes need to be connected with the asset, and additionally also a catalog of measures describing how to fix the damage. Freehand drawing of polygons got excellent feedback. This allows flexibility and is similar to what is also available in 2D-GIS systems today.

How accurate need the redlining symbols be placed. Symbols from a predefined palette do not need to be set precisely because the redlined objects will not be directly transferred into the GIS. An example would be placing a safety symbol indicating a safety area. In contrast when surveying a geospatial object or planning a new trench, the locations need to be determined precisely since the coordinates will be used to create an object in the GIS.

Spatial interaction. The spatial interaction method using the point-and-shoot metaphor for placing symbols in the geospatial model was rated high. Also, when marking a circular area on the ground, the variation of the radius with orientation changes got high scores. Salzburg AG workers were very enthusiastic about the interaction techniques and handling in general.

Overall advantage of redlining. As a major difference to conventional procedures, field workers mentioned plotted plans would be redundant for processes agreed before. The tablet PC systems they already use do not need paper prints, but only show data in 2D. Field workers also found it necessary to store and use large data sets with redlined objects. They mentioned that the main strength of the virtual redlining tool is the integration of functions for annotating geospatial objects and surveying of locations. The evaluations showed that the virtual redlining tool has the potential to improve the workflow of a variety of tasks such as planning, network inspection, network documentation and surveying street features.

7.5.4 Verification of abstract information

Field workers from E-Werk Gösting do not have a strong need to visualize such abstract information in the field, because they also use topographic data next to land-register data in their GIS. Before going on-site, they prepare and measure the land-register data in the main GIS in the office. For example, virtual semantic information, such as parcel areas can be visualized in a 2D GIS (see Figure 80). Employees from E-Werk Gösting rated 3D visualization most useful in places, where no borders can be seen, like in fields or woods. This verification tool was also rated helpful in a planning task when a new object, e.g. trench or pipe needs to be placed. In contrast, for Salzburg AG it is vital to determine and find land-register data in the real environment, since they do not use topological data in their GIS.

Generally, companies that only use land-register data can benefit from on-site visualizations. The visualization of parcel borders in an AR manner was found useful and timesaving by field workers from Salzburg AG and was rated to have a huge potential to simplify their work. Additionally, all field workers saw a huge potential for time savings using the AR style visualization for presenting such abstract information.

Field workers from E-Werk Gösting automatically rate a location as inaccurate, if only data of land registers and no topographical data is available. In this case they use GPS equipment to verify the cadastral data on-site. Since Salzburg AG only uses cadastral data, they need to perform this time consuming task every time to achieve accurate location information about a geospatial object. Salzburg AG saw an application area of the verification tool to determine parcel boundaries or parcel areas in particular for construction and excavation tasks.

7.6   Role of AR in field information systems

(Hammad, Garrett, & Karimi, 2002) investigated the potential and limitations of mobile AR for infrastructure field tasks. They identified that during construction, inspection, maintenance, and repair of transportation infrastructure projects, field engineer frequently refer to maps and other technical documents. Mobile AR has the potential for not only allowing users to automatically retrieve information in real-time based on their location but also display this information as augmentation to the view of the surrounding objects, such as roads, brigades, and tunnels. (Hammad, Garrett, & Karimi, 2002) identified the need for a hybrid tracking system as well as the need for developing distributed, topologically structured 3D GIS/CAD databases as a major step towards the practical implementation. Architecture, construction and civil engineering are generally recognized as one of the most applicable fields for VR/AR technology. The planning of new buildings involves important decision making on expensive matters, as well as communication and collaboration between various interest groups, which all in different ways signify the importance of having the future plans realistically presented to the stakeholders.

Let us take a look at a regularly performed work process or workflow in the utility industry. Typically field workers have the need to locate geospatial underground infrastructure objects during inspection tasks. Figure 81 (left) shows the workflow of an inspection task using a conventional approach based on 2D maps, whereas Figure 81 (right) shows the workflow using a 3-dimensional AR visualization. This suggests that using AR the workflow for an inspection task can be simplified. The following figures illustrate the difference between conventional maps and AR visualizations in an impressive way. Figure 82 shows a digital 2D plan showing underground infrastructure printed on paper.

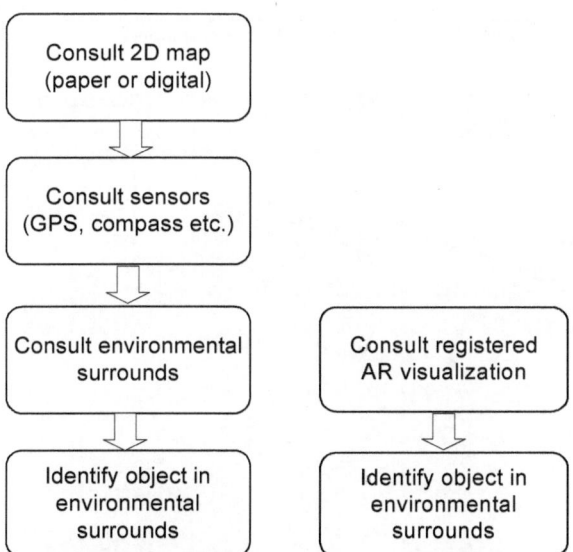

Figure 81: The workflow of an inspection task on the left follows the method used to identify an underground object (left) using a 2D map and (right) utilizing a 3-dimensional AR visualization.

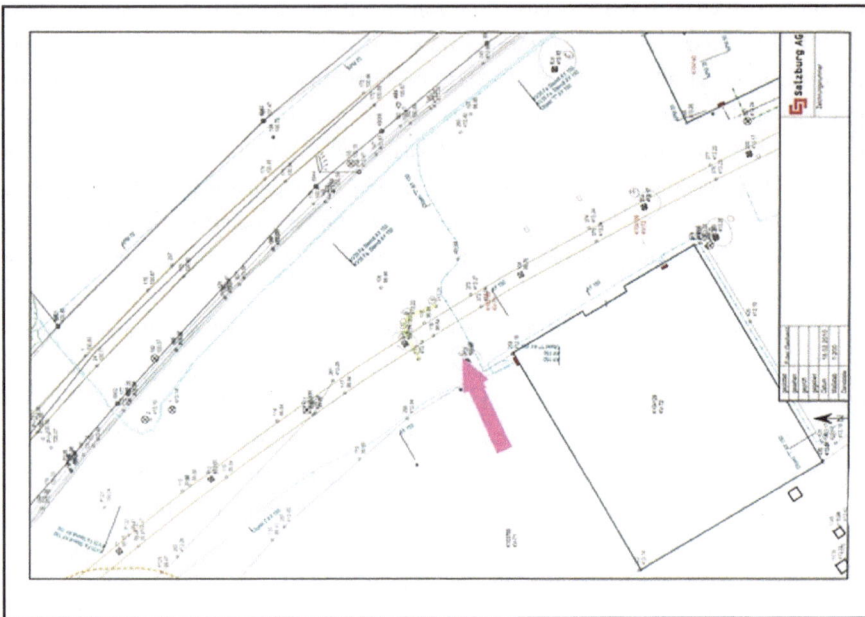

Figure 82: 2D map showing GIS features. User is located where the red arrow is pointing.

Figure 83: 3D AR visualization showing surveyed above ground features and subsurface features.

This is what field workers often take to the field for inspection tasks. In contrast, Figure 83 shows the corresponding GIS features while standing at the position indicated by the red arrow shown in Figure 82. The mobile AR user is oriented towards the direction of the arrow. Note that the position accuracy is very high since a RTK GPS receiver is used. The AR visualization using a trench along the pipes can convey depth and 3D information better than on a 2D map only.

A discussion and interview with test users from the utility sector following the field test took place to gain feedback on the performance of the 3D visualization. In a question-answer format the users were asked how well the 3D visualization displayed information, and whether the information was clear and easy to interpret. Like the initial questionnaire, the feedback was recorded on paper. The response indicates that the AR visualization helped the users to get a better and greater understanding of the surrounds. Moreover, in particular the task of identifying their position and orientation in the-field could be improved significantly.

The outdoor application presented here, provides the necessary research which leads to a fundamental platform for a 3rd generation field information system (1st generation: analog paper maps and plans, 2nd generation: digital maps on laptop computers of PDAs).

Using high-precision tracking of the mobile field worker this system aims to allow executing "key-hole surgery" on the underground infrastructure operated by enterprises of the utility sector. Such a development is expected to require less excavations; limited to the necessary, and thus offering faster execution of field work, reduced interruption to traffic and consequently less impact on the environment.

7.7 Discussion

If AR is to be taken seriously, tracking needs to fulfill user demands. The developed tracking approaches could achieve a quality of registration in 3D that was required by real-world users. The requirement for the position accuracy was to be more accurate than 30 centimeters. Furthermore, observations on the practical applicability of the presented applications, tools and interaction techniques were presented.

This chapter has also shown some of the possibilities that a mobile AR system has in civil engineering. Clearly, there are various advantages over a conventional 2D representation, among them more realistic presentation of geospatial object, automatic map scale and orientation, interaction and annotation possibilities. To show another example, take a look at Figure 84 that presents an AR view where the reference surface features are occluded by winter snow and are invisible to the user.

Field workers trying to locate features covered by winter snow would have a hard time in succeeding in this task. In line with hypothesis H4 stating that a 3D AR interface has advantages over conventional 2D maps in industrial outdoor settings, there is strong evidence for that. Especially in terms of workflow improvements, the 3D user interface shows advantages over a pure 2D interface. Considering the results from the evaluations of the AR prototypes, 3D user interfaces emerged as useful extensions to existing interfaces with a realistic potential for improving business processes in civil engineering.

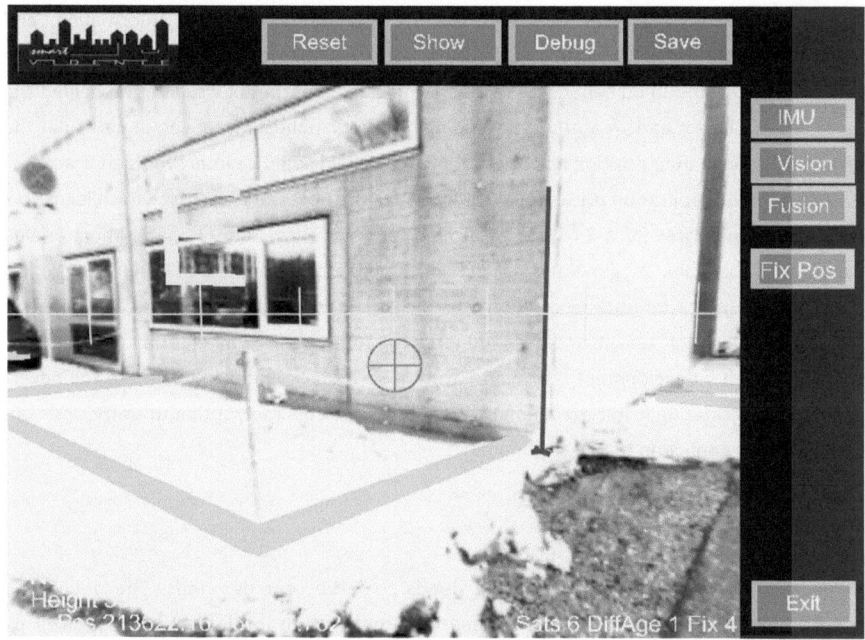

Figure 84: AR view with superimposed enclosures and base point of the building corner registered in 3D. The surface reference features are covered by winter snow.

Computing is not about computers
anymore. It is about living.

Nicholas Negroponte
Co-founder and director of the
MIT Media Laboratory

8. Conclusions

8.1 Reflection

In this thesis a variety of issues have been presented that illuminate issues in developing a state of the art mobile AR system. Mobile AR has outgrown its infancy and is ready
for early commercial deployment. This thesis presented a series of platforms, hybrid
tracking methods, geospatial modeling approaches and AR application prototypes to
assess the feasibility of key technologies in mobile AR. Investigating all these tasks is
very fascinating work. Next, a look is taken at the hypotheses formulated at the beginning of the thesis.

Hypothesis 1 of this thesis stated that AR needs special models that are different from existing virtual models, in particular because they need to support both visualization and tracking. The simple indoor navigation example at the beginning of Chapter 4 illustrated that the
semantic information of the model is used to assist the user in the task, e.g. re-calculating the
path through a building. Moreover, the example demonstrated that the geometric information of the model is not only used for visualization purposes but also for tracking fiducial
markers as the application derives the tracking data from the BAUML model. This example
helped creating awareness for the complex structure needed for AR models. Built on these
insights the transcoding pipeline was designed and implemented. This strongly contributed to
an automatic generation of AR models from existing legacy data sources. By containing visual
as well as non-visual information the resulting models support both the tracking subsystem
and the visualization subsystem.

Hypothesis 2 of the thesis stated that global referenced 3D models for augmented reality
can be created efficiently by using surveying procedures or by using legacy data. The transcoding approach demonstrated that models for AR can be created automatically and efficiently. One of the main strengths of this approach is the possibility of using procedural models. The advantage is that complex objects can be represented with a small number of parameters, thus the description is very compact and would allow very short download times via a

wireless link. Furthermore, the visualization of the procedural models can be very detailed and impressive but this also affects the rendering performance. Consequently, depending on the application's needs, specific objects can be represented as procedural models and others can be represented as simple Open Inventor models. Using this approach, 3D models cannot only be generated efficiently; they even more contribute to more efficiency for the visualization and tracking subsystem.

Improvements in model creation could increase the performance and enhance the capabilities of AR applications, and reduce the cost of their development. Model creation is critical for AR models for use in AR can be sourced in a number of ways: pre-existing models can be re-purposed in some cases, models can be created in advance for the specific application, or models can be created by the AR system on the fly based on environment sensing. For some uses, such as in medical and industrial applications, models must be very precise. In other cases, such as games and applications used on mobile phones, high precision often is not required.

Simultaneous to model creation, tracking the user's or device's pose is of importance to achieve the level of robustness, accuracy and stability required by the application. Whereas hypotheses 1 and 2 focused on modeling for AR, hypothesis 3 claimed that the accuracy of hybrid tracking can be sufficiently high for civil engineering tasks in an outdoor environment. For example, field workers demand a positional accuracy of better than 30 centimeters for inspection or maintenance tasks. Even more, for surveying tasks the position must be accurate to a few centimeters. These demands are pretty tough and hard to achieve considering also form factors and weight restrictions for AR hardware setups. Experiments proved that position accuracies fulfilling these requirements can be achieved, given that at least four satellites are visible to the differential GPS receiver. Extending highly accurate position estimation into urban canyons still reflects a research challenge. Furthermore, orientation estimation is critical because magnetic compasses tend to deviate in the presence of electromagnetic fields. By applying a hybrid tracking approach combining both magnetic and vision-based tracking, a more robust and stable orientation estimate could be calculated. Vision-based tracking promises to improve pose tracking in urban canyons by using accurate urban 3D models (model-based tracking). In this case, semantic geospatial models are key enablers supporting vision-based tracking.

All these steps were necessary — namely generating geospatial urban 3D models, sufficiently accurate tracking and building the according AR hardware setups — in order to create the presented AR applications. If any of these steps would be left away, it would not be possible to create a working AR application.

Anyway, assuming that these key ingredients for a mobile outdoor AR application are available, there is still an important question to ask. The question is, if AR interfaces can provide a significant improvement over existing interfaces while performing specific tasks in outdoor industrial settings. The collaboration with various organizations and real-world users had constructive influences in steering the development of the AR prototypes. When observing current developments in industry, we are in line with the trend towards mobile devices as well as the desire to present data more true to their nature. For the latter issue, AR has the huge potential to provide an intuitive user interface and claim its permanent existence. In line with Hypothesis H4 stating that a 3D AR interface has advantages over conventional 2D maps in outdoor industrial settings, there is strong evidence for that according to the evaluation results. In particular, considering the workflow improvements, a 3D user interface shows advantages over a pure 2D interface.

Many civil engineering tasks require to access geospatial data in the field and reference the stored information to the real world situation. Augmented Reality (AR), which interactively overlays 3D graphical content directly over a view of the world, can be a useful tool to visualize but also create, edit and update geospatial data representing real world artifacts. In two papers published after this dissertation I present research results on the next-generation field information system for companies relying on geospatial data, providing mobile workforces with capabilities for on-site inspection and planning, data capture and as-built surveying. To achieve this aim, we used mobile AR technology for on-site surveying of geometric and semantic attributes of geospatial 3D models on the user's handheld device. The interactive 3D visualizations automatically generated from production databases provide immediate visual feedback for many tasks and lead to a round-trip workflow where planned data is used as a basis for as-built surveying through manipulation of the planned data. Classically, surveying of geospatial objects is a typical scenario performed from utility companies on a daily basis. We demonstrate a mobile AR system that is capable of these operations and present first field trials with expert end-users from utility companies. The initial results show that the workflows of planning and surveying of geospatial objects benefit from our AR approach. This work advances the approach presented in this book and is published in Schall, Zollmann and Reitmayr (2012) and in Zollmann et al. (2012).

8.2   Road map

A few more years of technological advancement and industry maturity is required before we start seeing things that will have a lasting effect on our daily lives. The author is
certain that in the years to come, further advancements in the state of the art in mobile
AR will happen as the field is becoming even more interdisciplinary. Some of the following issues will likely be addressed.

The "big picture" vision for AR is simple: the world as user interface – using real objects,
places and people encountered in the world as reference points for additional computer-
generated information. Computing will become smaller and almost unnoticeable and be part
of nearly every aspect of our lives. Future mobile devices will further push AR forward. For
now, users are holding up their mobile devices and often have to interact with small screens.
In the future, users will simply walk into a room and the device will know where they are.
Users will always have directions to get where they need to be without having to look it up.
Everybody will feel the effects of how mobile, ubiquitous computing, augmented reality,
smart devices, embedded sensors and automation changes the daily lifes.

Moreover, it is a notable challenge for GIS to adapt a new user interface with augmented
reality. However the obvious trend in the industrial environment is the emphasis on both
mobility and real-time data capabilities. In order to realize GIS on mobile augmented reality
displays, a standard data format needs to be established. This will probably emerge when
industry leaders push the concept forward. Based on this, there is a need for a strong content
creation pipeline.

Mobile mapping refers to a means of collecting geospatial data using mapping sensors
that are mounted on a mobile platform. Direct geo-referencing of digital image sequences is
accomplished through the use of navigation and positioning techniques. The research on
mobile mapping dates back to the late 1980s. Looking back, after using aerial images from
planes for mapping the physical surface of our earth, already dedicated cars are equipped
with sensors to perform mobile mapping at street level. In future, handheld mobile mapping
systems will be used for mapping even smaller structures than streets. This could culminate
in a very detailed model of many human environments, a so called *Virtual Habitat*. At the
same time, image recognition and visual tracking algorithms could use these models for pose
estimation. In this way, accurate semantic 3D models can greatly contribute to more accurate
tracking solutions in urban environments.

Appendix

Questionnaire

The following four pages contain the questionnaire handed out for the evaluation of the Vesp`R setup performed in September 2007.

Subject number

Gender   female | male

Hand size   small | medium | large

Dexterity   left-handed | right-handed | ambidextrous

Please answer the following questions by putting in cross in the „O"

---

1. How do you rate the **overall weight** of the device

Too heavy   Mediocre / OK   Light enough

O -------- O -------- O -------- O -------- O -------- O -------- O

2. How do you rate the **weight balance** of the device

Very bad   Mediocre / OK   Very good

O -------- O -------- O -------- O -------- O -------- O -------- O

3. How do you rate the **ergonomics of the grip** of the „joysticks"

Very bad   Mediocre / OK   Very good

O -------- O -------- O -------- O -------- O -------- O -------- O

→ Some more questions on the next page!

4. How do you rate the **grip material** of the „joysticks" (the black rubber)

Very bad   Mediocre / OK   Very good

O ------- O ------- O ------- O ------- O ------- O ------- O

5. Did you notice any **fatigue** („tiring reaction") caused by using the device?

Very tiring   Mediocre / OK   Not tiring at
all

O ------- O ------- O ------- O ------- O ------- O ------- O

6. How do you rate the **placement of the controllers** (buttons or minijoystick) , i.e. could you easily reach and control them?

Very bad   Mediocre / OK   Very good

O ------- O ------- O ------- O ------- O ------- O ------- O

7. Did you often **switch between focal planes**, i.e. did you often change between looking at the screen and looking at the environment itself you were observing?

Very often   Sometimes   Not at all

O ------- O ------- O ------- O ------- O ------- O ------- O

8. Did it bother you to **switch between focal planes**?

Very often   Sometimes   Not at all

O ------- O ------- O ------- O ------- O ------- O ------- O

→ some more question to go!

9. How **effective** could you control the application (interaction) ?

Very hard to control   OK   Very easy to
control

| | | | | |

O -------- O -------- O -------- O -------- O -------- O -------- O

10. How do you rate the „**coolness**" of this device?

Dull   OK   Way cool!

| | | | | |

O -------- O -------- O -------- O -------- O -------- O -------- O

11. How do you rate the **usefulness of the application** (showing surface and subsurface structures)?

Not usefull   OK   Very useful

| | | | | |

O -------- O -------- O -------- O -------- O -------- O -------- O

12. How do you rate the quality of the interaction technique „**magic lens**"?

Very bad   OK   Very good

| | | | | |

O -------- O -------- O -------- O -------- O -------- O -------- O

13. How do you rate the quality of the interaction technique „**x-ray selection**"?

Very bad   OK   Very good

| | | | | |

O -------- O -------- O -------- O -------- O -------- O -------- O

Miscelaneous comments:
(Missing items, problems etc.)
....
....
....

Thanks for your participation!

Observation questions

User ID	Hand trembling / fatigue			Holding device			Special grip?	Amount of interaction (muscular activity)		
	much	normal	little	very low	middle	eye-height		low	normal	high

Centimeter grid

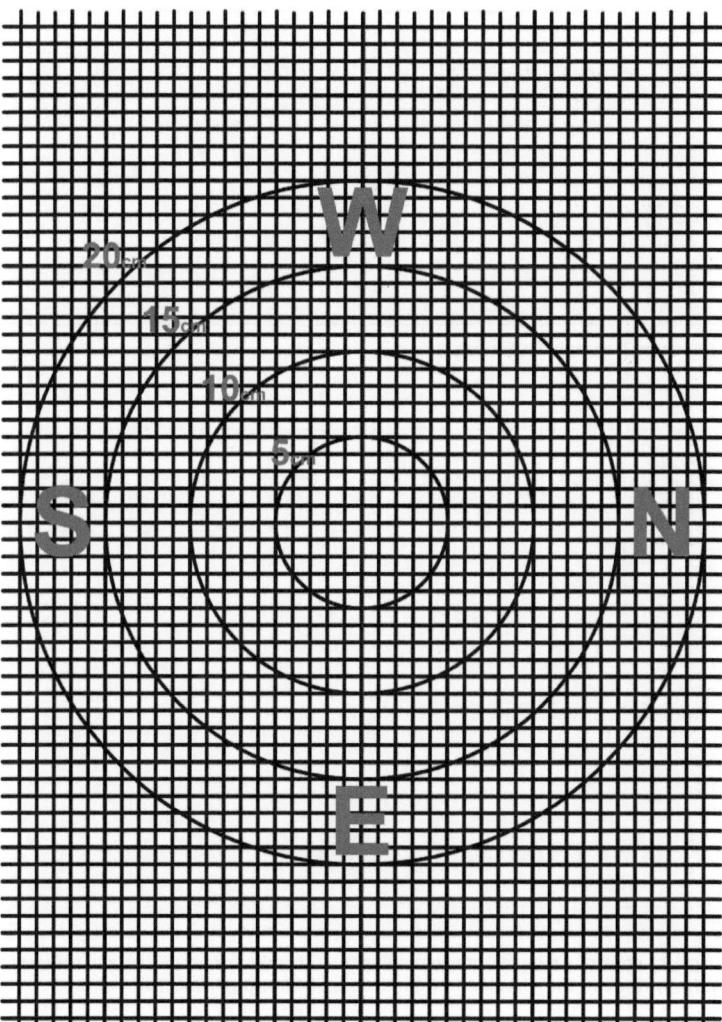

Figure 85: Centimetre grid used for accuracy experiments.

Transcoding output nodes

```
<gml:featureMember>
    <Feature id="123" name="name" alias="Der Name" group="gruppenname" groupAlias="Gruppen-Name">
        <property name="eigenschaft1" alias="Eigenschaft 1" type="typ"></property>
        ...
        ...
        <geometry>
            <gml:LineString>
                <gml:posList>
                    1 2 3 11 12 13
                </gml:posList>
            </gml:LineString>
        </geometry>
    </Feature>
</gml:featureMember>
```

Figure 86: Feature node in GeographyML format.

```
SoSeparator
{
        SoStyleProperty
        {
                keys    ["id", "name", "alias", "group", "groupAlias", "level", "parent",
"aussenschutz_alias", "aussenschutz_type", "aussenschutz_value", "bauleiter_alias"]

                values  ["11414423", "m_wa_la", "WA Leitungsabschnitt", "wasser", "Wasser", "", "",
"Außenschutz", "string", "unbekannt", "Bauleiter"]
        }

        SoStyledSubgraph
        {
                content SoMultiPipeKit
                {
                        caps FALSE
                        radius 0.150
                        numFaces 10
                        coords [   -3.968224 -18.068376 355.963040,  7.401852 -11.421171 355.819496,
9.070578 -10.438964 355.820909, 37.745911 6.429745 355.699489]
                        lineIndices [ 4 ]
                }
        }
}
```

Figure 87: Open Inventor node of a pipe feature containing the subnodes for properties and geometry. The geometry node only contains the coordinates from the original GeographyML file.

```
DEF SEP SoSeparator
{
        SoStyledSubgraph
        {
                content SoMultiPipeKit
                {
                        caps       FALSE
                        radius     0.1
                        numFaces   10
                        coords [0 9 0, 0 2 0, 0.0979 1.382 0, 0.382 0.8244 0, 0.8244 0.382 0, 1.382
0.0979 0, 2 0 0, 7 0 0, 7.7874 -0.0329 0, 8.5693 -0.1316 0, 9.3402 -0.2952 0, 10.0947 -0.5226 0, 18
-4 0]

                        lineIndices [13]

                        #!#ShapeGML-TRANSFORMATION DATA#!#
                        #!#IsTrasse#FALSE#
                        #!#TrasseRad#0#
                        #!#RingEnable#FALSE#
                        #!#RingThick#0#
                        #!#RingLength#0#
                        #!#RingMaterial#std8hi#
                        #!#TubeThick#0#
                        #!#OpenAngle#0#
                        #!#CurveRad#2#
                        #!#CurveStart#0#
                        #!#CurveSteps#5#
                        #!#OrigCoords#0  9 0,  0 0 0,  9 0 0,  18 -4 0#
                        #!#Offset#0 0 0#
                }
        }
}
```

Figure 88: Open Inventor node of a pipe feature containing the subnodes for properties and geometry. The geometry node contains the GenerativeML parameters.

Bibliography

Abawajy, J. H. (2009). Advances in pervasive computing: GUEST EDITORIAL. *International Journal of Pervasive Computing and Communications*, *5*(1), 4-8. doi: 10.1108/17427370910950276.

APOS - Austrian Positioning Service. (2010). Retrieved November 23, 2010, from http://www.bev.gv.at/portal/page?_pageid=713,1571538&_dad=portal&_schema=POR TAL.

Azuma, R., (1997). A survey of augmented reality. *Presence-Teleoperators and Virtual Environments* (Vol. 6, p. 355–385).

Azuma, R., Hoff, B., Neely III, H., & Sarfaty, R. (2002). A motion-stabilized outdoor augmented reality system. *Virtual Reality, 1999. Proceedings., IEEE* (p. 252–259). IEEE. Retrieved November 26, 2010, from http://ieeexplore.ieee.org/xpls/abs_all.jsp ?arnumber=756959.

Azuma, R., Weon Lee, J., Jiang, B., Park, J., You, S., & Neumann, U. (1999). Tracking in unprepared environments for augmented reality systems. *Computers & Graphics*, *23*(6), 787–793. Elsevier. Retrieved November 26, 2010, from http://linkinghub.elsevier.com/ retrieve/pii/S0097849399001041.

Bachmann, E., Duman, I., Usta, U., McGhee, R., Yun, X., & Zyda, M. (2002). Orientation tracking for humans and robots using inertial sensors. *Computational Intelligence in Robotics and Automation, 1999. CIRA'99. Proceedings. 1999 IEEE International Symposium on* (p. 187–194). IEEE. Retrieved November 26, 2010, from http://ieeexplore.ieee.org/xpls/ abs_all.jsp?arnumber=810047.

Badard, T. (2006). Geospatial service oriented architectures for mobile augmented reality. *Proc. of the 1st International Workshop on Mobile Geospatial Augmented Reality* (p. 73–77). Citeseer. Retrieved November 29, 2010, from http://citeseerx.ist.psu.edu/ viewdoc/download?doi=10.1.1.126.6364&rep=rep1&type=pdf.

Bane, R., & Höllerer, T. (2004). Interactive tools for virtual x-ray vision in mobile augmented reality. *Third IEEE and ACM International Symposium on Mixed and Augmented Reality, 2004. ISMAR 2004.* (p. 231–239). Arlington, VA.

Bleser, G., Wuest, H., & Strieker, D. (2007). Online camera pose estimation in partially known and dynamic scenes. *Mixed and Augmented Reality, 2006. ISMAR 2006. IEEE/ACM International Symposium on* (p. 56–65).

Breuss-Schneeweis, P. (2009). An AR Travel Guide. Retrieved from http://www.mobilizy. com/wikitude.php.

Bruenig, M., & Zlatanova, S. (2006). 3D Geo-DBMS. *Directions Magazine*. Retrieved from http://www.directionsmag.com/printer.php?article_id=694.

Caruso, M. J. (1998). Applications of magnetoresistive sensors in navigation systems. *PROGRESS IN TECHNOLOGY* (Vol. 72, p. 159–168). SAE INTERNATIONAL. Retrieved November 23, 2010, from http://www.ssec.honeywell.com/position-sensors /datasheets/ sae.pdf.

DiVerdi, S., Wither, J., & Höllerer, T. (2008). Envisor: Online environment map construction for mixed reality. *Virtual Reality Conference, 2008. VR'08. IEEE* (p. 19–26). IEEE. Retrieved November 23, 2010, from http://ieeexplore.ieee.org/xpls/abs_all.jsp ?arnumber=4480745.

EPOSA - Echtzeit Positionierung Austria. (2010). . Retrieved November 26, 2010, from http://www.eposa.at.

Feiner, S., MacIntyre, B., Höllerer, T., & Webster, A. (1997). A touring machine: Prototyping 3D mobile augmented reality systems for exploring the urban environment. *Personal and Ubiquitous Computing* (Vol. 1, p. 208–217). Springer. Retrieved November 23, 2010, from http://www.springerlink.com/index/t553412326742258.pdf.

Fitzmaurice, G., & Buxton, W. (1994). The chameleon: Spatially aware palmtop computers. *Conference companion on Human factors in computing systems* (p. 451–452). New York, NY, USA: ACM. Retrieved November 23, 2010, from http://portal.acm.org/citation.cfm ?id=259963.260460&type=series.

Fröhlich, P. (2009). Mobile spatial interaction. *Personal and Ubiquitous Computing*. Retrieved November 24, 2010, from http://portal.acm.org/citation.cfm?id=1527370.

Geography Markup Language | OGC®. (2010). Retrieved November 23, 2010, from http://www.opengeospatial.org/standards/gml.

Groves, P. D. (2008). Principles of GNSS, inertial, and multi-sensor integrated navigation systems. Boston London: Artech House.

Hallaway, D. (2004). Bridging the gaps: Hybrid tracking for adaptive mobile augmented reality. *Applied Artificial Intelligence*.

Hammad, A., Garrett, J. H., & Karimi, H. A. (2002). Potential of mobile augmented reality for infrastructure field tasks. *7th International Conference on 3D Web Technology: Tempe, AZ, USA*.

Havemann, S., & Fellner, D. (2004). Generative parametric design of gothic window tracery. IEEE Computer Society. Retrieved November 29, 2010, from http://www.computer.org /portal/web/csdl/doi/10.1109/SMI.2004.1314525.

Hill, A., MacIntyre, B., & Gandy, M. (2010). KHARMA: A KML/HTML Architecture for Mobile Augmented Reality Applications.

Hofmann-Wellenhof, B., Legat, K., & Weiser, M. (2004). Navigation Principles of Positioning and Guidance. *Springer, Wien/New York* (Vol. 85). Wien/New York: Springer.

Hu, X., Liu, Y., Wang, Y., Hu, Y., & Yan, D. (2005). Autocalibration of an electronic compass for augmented reality. *Proceedings of the 4th IEEE/ACM International Symposium on Mixed and Augmented Reality* (p. 182–183).

Huisman, O., & Forer, P. (2008). Progress in the application of Time-Geographic concepts to urban micro-process. *GISSCIENCE*. Park City, Utah, USA.

Höllerer, T., Feiner, S., Terauchi, T., Rashid, G., & Hallaway, D. (1999). Exploring MARS: developing indoor and outdoor user interfaces to a mobile augmented reality system. *COMPUT GRAPHICS(PERGAMON)* (Vol. 23, p. 779–785).

Junghanns, S., Ranzinger, M., Schall, G., & Reitmayr, G. (2010). Technical report, *Specification VIDENTEGML V1.0.2*. Graz University of Technology

Junghanns, S., Schall, G., & Schmalstieg, D. (2008). VIDENTE – What lies beneath? A new approach of locating and identifying buried utility assets on site. *Proceedings of the 5th International Symposium on LBS & TeleCartography (LBS'08), showcase*. Salzburg, Austria.

Kalkofen, D., Mendez, E., & Schmalstieg, D. (2007). Interactive focus and context visualization for augmented reality. *Proceedings of 6th IEEE and ACM International Symposium on Mixed and Augmented Reality, 2007. ISMAR 2007.* (p. 191–201). IEEE. Retrieved November 23, 2010, from http://ieeexplore.ieee.org/xpls/abs_all.jsp?arnumber= 4538846.

Kato, H., Billinghurst, M., Blanding, B., & May, R. (1999). ARToolKit. *Hiroshima City University*.

King, G. R., Piekarski, W., & Thomas, B. H. (2005). ARVino-outdoor augmented reality visualisation of viticulture GIS data. *Proceedings of Fourth IEEE and ACM International Symposium on Mixed and Augmented Reality, 2005.* (p. 52–55). IEEE. Retrieved November 23, 2010, from http://ieeexplore.ieee.org/xpls/abs_all.jsp?arnumber= 1544663.

Klein, G., & Drummond, T. (2004). Sensor fusion and occlusion refinement for tablet-based ar. *Proceedings of the 3rd IEEE/ACM International Symposium on Mixed and Augmented Reality* (p. 38–47).

Klein, G., & Murray, D. (2009). Parallel tracking and mapping on a camera phone. *Mixed and Augmented Reality, 2009. ISMAR 2009. 8th IEEE International Symposium on* (p. 83–86). IEEE. Retrieved November 28, 2010, from http://ieeexplore.ieee.org/xpls/abs_all.jsp?arnumber=5336495.

Kolbe, T. H., Gröger, G., & Pluemer, L. (2008). CityGML–3D city models and their potential for emergency response. *Geospatial Information Technology for Emergency Response* (p. 257). Taylor & Francis Group.

Kruijff, E., & Veas, E. (2007). Vesp'R-Transforming Handheld Augmented Reality. *Proceedings of the 6th IEEE and ACM International Symposium on Mixed and Augmented Reality* (p. 1–2). Japan.

La Beaujardiere, J. de. (2004). OpenGIS Web Map Service (WMS) Implementation Specification. *WWW document, http://portal. opengeospatial. org/files*.

Lowe, D. G. (2004). Distinctive image features from scale-invariant keypoints. *International journal of computer vision* (Vol. 60, p. 91–110). Springer.

Matas, J., Chum, O., Urban, M., & Pajdla, T. (2004). Robust wide-baseline stereo from maximally stable extremal regions. *Image and Vision Computing* (Vol. 22, p. 761–767). Elsevier. Retrieved November 23, 2010, from http://linkinghub.elsevier.com/ retrieve/pii/S0262885604000435.

Mendez, E., Schall, G., Havemann, S., Fellner, D., Schmalstieg, D., & Junghanns, S. (2008). Generating Semantic 3D Models of Underground Infrastructure. *IEEE Computer Graphics and Applications* (Vol. 28, pp. 48-57). doi: 10.1109/MCG.2008.53.

Milgram, P., & Kishino, F. (1994). A taxonomy of mixed reality visual displays. *IEICE Transactions on Information and Systems E series D* (Vol. 77, p. 1321–1321). Citeseer. Retrieved November 23, 2010, from http://citeseerx.ist.psu.edu/viewdoc/download ?doi= 10.1.1.102.4646&rep=rep1&type=pdf.

Newman, J., Bornik, A., Pustka, D., Echtler, F., Huber, M., Schmalstieg, D., et al. (2007). Tracking for distributed mixed reality environments. *Workshop on Trands and Issues in Tracking for Virtual Environments at the IEEE Virtual Reality Conference (VR'07)*. Charlotte, NC, USA.

Newman, J., Fraundorfer, F., Schall, G., & Schmalstieg, D. (2005). Construction and Maintenance of Augmented Reality Environments using a Mixture of Autonomous and Manual Surveying Techniques. *Proceedings of 7th conference on Optical 3-D Measurement Techniques*. Vienna, Austria.

Newman, J., Ingram, D., & Hopper, A. (2001). Augmented reality in a wide area sentient environment. *Proceedings. IEEE and ACM International Symposium on Augmented Reality* (p. 77–86). New York, NY, USA.

Newman, J., Schall, G., Barakonyi, I., Schürzinger, A., & Schmalstieg, D. (2006). Wide-Area Tracking Tools For Augmented Reality. *Advances In Pervasive Computing*. Dublin, Ireland.

Newman, J., Schall, G., & Schmalstieg, D. (2006). Modelling and Handling Seams in Wide-Area Sensor Networks. *2006 10th IEEE International Symposium on Wearable Computers* (pp. 51-54). IEEE. doi: 10.1109/ISWC.2006.286342.

Nistér, D. (2004). An efficient solution to the five-point relative pose problem. *Pattern Analysis and Machine Intelligence, IEEE Transactions on, 26*(6), 756–770. IEEE. Retrieved November 23, 2010, from

Olsen Jr, D. R. (2007). Evaluating user interface systems research. *Proceedings of the 20th annual ACM symposium on User interface software and technology* (p. 251–258). ACM. Retrieved December 4, 2010, from http://portal.acm.org/citation.cfm?id=1294256.

Oxford Dictionaries Online - English Dictionary and Language Reference. (2010). . Retrieved November 30, 2010, from http://oxforddictionaries.com/?attempted=true.

Paelke, V., & Brenner, C. (2007). Development of a mixed reality device for interactive on-site geo-visualization. *Proceedings of 18th Simulation and Visualization Conference* (pp. 237-248).

Piekarski, W., & Thomas, B. H. (2001). Tinmith-metro: New outdoor techniques for creating city models with an augmented reality wearable computer. *Proceedings of the International Symposium on Wearable Computers (ISWC´01)* (p. 31). Zurich, Switzerland: Published by the IEEE Computer Society.

Priestnall, G., & Polmear, G. (2006). Landscape Visualisation: From lab to field. *Proceedings of the First International workshop on mobile geospatial augmented reality* (p. 29–20).

Reitmayr, G., & Drummond, T. W. (2006). Going out: robust model-based tracking for outdoor augmented reality. *IEEE/ACM International Symposium on Mixed and Augmented Reality, 2006. ISMAR 2006.* (p. 109–118). Santa Barbara, USA.

Reitmayr, G., & Drummond, T. W. (2007). Initialisation for visual tracking in urban environments. *6th IEEE and ACM International Symposium on Mixed and Augmented Reality, 2007. ISMAR 2007.* (p. 161–172). Nara, Japan.

Reitmayr, G., & Schmalstieg, D. (2001). Opentracker-an open software architecture for reconfigurable tracking based on XML. *Proceedings. IEEE Virtual Reality, 2001.* (p. 285–286). IEEE. Retrieved November 23, 2010, from http://ieeexplore.ieee.org/xpls/abs_all.jsp?arnumber=913799.

Reitmayr, G., & Schmalstieg, D. (2004). Collaborative augmented reality for outdoor navigation and information browsing. *Symposium Location Based Services and TeleCartography* (p. 31–41). Vienna, Austria.

Rekimoto, J. (2001). Navicam: A palmtop device approach to augmented reality. *Fundamentals of wearable computers and augmented reality* (p. 353). CRC.

Rekimoto, J. (2002). Matrix: A realtime object identification and registration method for augmented reality. *Computer Human Interaction, 1998. Proceedings. 3rd Asia Pacific* (p. 63–68). IEEE. Retrieved November 28, 2010, from http://ieeexplore. ieee.org/xpls/abs_all.jsp?arnumber=704151.

Ribo, M., Lang, P., Ganster, H., Brandner, M., Stock, C., & Pinz, A. (2002). Hybrid tracking for outdoor augmented reality applications. *IEEE Computer Graphics and Applications*, 54–63. Published by the IEEE Computer Society.

Roberts, G. W., Evans, A., Dodson, A., Denby, B., Cooper, S., & Hollands, R., others. (2002). The use of augmented reality, GPS and INS for subsurface data visualization. *FIG XXII International Congress of the FIT*. Washington DC, USA.

Robertson, C. M., MacIntyre, B., & Walker, B. N. (2008). An evaluation of graphical context when the graphics are outside of the task area. *Mixed and Augmented Reality, 2008. ISMAR 2008. 7th IEEE/ACM International Symposium on* (p. 73–76).

Robinson, A. H., & Petchenik, B. B. (1976). *The nature of maps*. University of Chicago Press. Retrieved December 5, 2010, from http://drc.ohiolink.edu/handle/2374.OX/21219.

Rohs, M., Schöning, J., Krueuger, A., & Hecht, B. (2007). Towards real-time markerless tracking of magic lenses on paper maps. *Adjunct Proceedings of the 5th Intl. Conference on Pervasive Computing (Pervasive), Late Breaking Results* (p. 69–72).

Schall, G., Grabner, H., Grabner, M., Wohlhart, P., Schmalstieg, D., & Bischof, H. (2008a). 3D tracking in unknown environments using on-line keypoint learning for mobile augmented reality. *2008 IEEE Computer Society Conference on Computer Vision and Pattern Recognition Workshops* (pp. 1-8). IEEE. doi: 10.1109/CVPRW.2008.4563134.

Schall, G., Junghanns, S., & Schmalstieg, D. (2008a). The Transcoding Pipeline: Automatic Generation of 3D Models from Geospatial Data Sources. *Workshop on Trends in Pervasive and Ubiquitous Geotechnology and Geoinformation in conj. with the 5th International Conference on GIScience (GISCIENCE 2008)*. Park City, Utah, USA.

Schall, G., Junghanns, S., & Schmalstieg, D. (2010). VIDENTE - 3D Visualization of Underground Infrastructure using Handheld Augmented Reality. *Geohydroinformatics - Integrating GIS and Water Engineering*.

Schall, G., Mendez, E., & Kruijff, E. (2008b). Handheld augmented reality for underground infrastructure visualization. *Personal and Ubiquitous Computing, Springer*.

Schall, G., Mendez, E., & Schmalstieg, D. (2008c). Virtual redlining for civil engineering in real environments. *Proceedings of the 7th IEEE/ACM International Symposium on Mixed and Augmented Reality* (pp. 95-98).

Schall, G., Mulloni, A., & Reitmayr, G. (2010). North-centred Orientation Tracking on Mobile Phones. *IEEE Int. Symposium on Mixed and Augmented Reality 2010 (ISMAR´10).* Seoul, South Korea,.

Schall, G., Newman, J., & Schmalstieg, D. (2005). Rapid and accurate deployment of fiducial markers for augmented reality. *Proceedings 10th Computer Vision Winter Vision Workshop.* Zell an der Pram, Upper Austria.

Schall, G., & Schmalstieg, D. (2008). Interactive Urban Models generated from Context-Preserving Transcoding of Real-Wold Data. *5th International Conference on GIScience, abstracts volume.* Park City, Utah, USA.

Schall, G., Wagner, D., Reitmayr, G., Taichmann, E., Wieser, M., Schmalstieg, D., Hoffmann-Wellenhof B. (2009). Global pose estimation using multi-sensor fusion for outdoor Augmented Reality. *2009 8th IEEE International Symposium on Mixed and Augmented Reality* (pp. 153-162). IEEE.

Schall, G., Zollmann S., Reitmayr G. (2012). Smart Vidente: advances in mobile augmented reality for interactive visualization of underground infrastructure. Personal and Ubiquitous Computing, Theme issue on Interaction and Visualization of 3D Virtual Environments on Mobile Devices, Guest Editors: José M. Noguera, Juan Carlos Torres. Springer. doi:10.1007/s00779-012-0599-x

Schmalstieg, D., Fuhrmann, A., Hesina, G., Szalavári, Z., Encarnaçao, L. M., Gervautz, M., et al. (2002). The studierstube augmented reality project. *Presence: Teleoperators & Virtual Environments* (Vol. 11, p. 33–54). MIT Press.

Schmalstieg, D., & Reitmayr, G. (2006). Augmented Reality as a Medium for Cartography. Springer-Verlag.

Schmalstieg, D., Schall, G., Wagner, D., Barakonyi, I., Reitmayr, G., Newman, J., et al. (2007). Managing Complex Augmented Reality Models. *IEEE Computer Graphics and Applications* (Vol. 27, pp. 48-57). doi: 10.1109/MCG.2007.85.

Steggles, P., & Gschwind, S. (2005). The Ubisense Smart Space Platform. *Advances In Pervasive Computing, Adjunct Proceedings Of The Third International Conference On Pervasive Computing. Vol. 191.*

Strauss, P. S., & Carey, R. (1992). An object-oriented 3D graphics toolkit. *ACM SIGGRAPH Computer Graphics* (Vol. 26, p. 341–349).

Sutherland, I. E. (1968). A head-mounted three dimensional display. *Proceedings of the December 9-11, 1968, fall joint computer conference, part I* (p. 757–764). ACM.

Thomas, B. H., & Piekarski, W. (2002). Glove based user interaction techniques for augmented reality in an outdoor environment. *Virtual Reality* (Vol. 6, p. 167–180). Springer.

Vretanos, P. (2002). Web Feature Service Implementation Specification. *OpenGIS project document: OGC*, 02–058. Retrieved November 29, 2010, from http://www.citeulike.org /user/ehjuerrens/article/6075076.

Wagner, D., Mulloni, A., Langlotz, T., & Schmalstieg, D. (2010). Real-time panoramic mapping and tracking on mobile phones. *Virtual Reality Conference (VR), 2010 IEEE* (p. 211–218).

Wagner, D., Pintaric, T., Ledermann, F., & Schmalstieg, D. (2005). Towards massively multi-user augmented reality on handheld devices. *Pervasive Computing* (p. 208–219). Springer.

Wagner, D., Reitmayr, G., Mulloni, A., Drummond, T., & Schmalstieg, D. (2008). Pose tracking from natural features on mobile phones. *7th IEEE/ACM International Symposium on Mixed and Augmented Reality, ISMAR 2008.* (p. 125–134). IEEE.

Wagner, D., & Schmalstieg, D. (2003). First steps towards handheld augmented reality. *Seventh IEEE International Symposium on Wearable Computers, 2003. Proceedings.* (pp. 127-135). IEEE. doi: 10.1109/ISWC.2003.1241402.

Wagner, D., & Schmalstieg, D. (2007). Artoolkitplus for pose tracking on mobile devices. *Proceedings of 12th Computer Vision Winter Workshop (CVWW'07)* (p. 139–146). Citeseer. Retrieved November 24, 2010, from http://citeseerx.ist.psu.edu/viewdoc/ download?doi=10.1.1.157.1879&rep=rep1&type=pdf.

Walczak, K., & Cellary, W. (2003). X-VRML for advanced virtual reality applications. *Computer* (Vol. 36, p. 89–92). IEEE. Retrieved November 23, 2010, from http://ieeexplore.ieee.org/xpls/abs_all.jsp?arnumber=1185226.

Weiser, M. (1991). The computer for the twenty-first century. *Scientific American* (Vol. 265, p. 94–104). Retrieved November 23, 2010, from http://www.citeulike.org/user/ karin/article/771482.

Welcome to the OGC Website | OGC®. (2010). Retrieved November 23, 2010, from http://www.opengeospatial.org/.

Whiteside, A., & Evans, J. (2006). Web coverage service (WCS) implementation specification. *Open Geospatial Consortium, Inc*, 129.

Wither, J., DiVerdi, S., & Höllerer, T. (2009). Annotation in outdoor augmented reality. *Computers & Graphics* (Vol. 33, p. 679–689). Elsevier.

You, S., Neumann, U., & Azuma, R. (2002). Hybrid inertial and vision tracking for augmented reality registration. *Virtual Reality, 1999. Proceedings., IEEE* (p. 260–267). IEEE. Retrieved November 28, 2010, from http://ieeexplore.ieee.org/xpls/abs_all.jsp?arnumber= 756960.

Zhang, X., & Gao, L. (2009). A novel auto-calibration method of the vector magnetometer. *Electronic Measurement & Instruments, 2009. ICEMI'09. 9th International Conference on* (p. 1). IEEE. Retrieved November 28, 2010, from http://ieeexplore.ieee.org/xpls/abs_all.jsp?arnumber=5274904.

Zollmann, S., Kalkofen, D., Mendez, E., & Reitmayr, G. (2010). Image-based Ghostings for Single Layer Occlusions in Augmented Reality. *IEEE/ACM International Symposium on Mixed and Augmented Reality. ISMAR 2010. Korea, Seoul.*

Zollmann, S., Schall, G., Junghans, S., Reitmayr, G. (2012) Comprehensible and Interactive Visualizations of GIS Data in Augmented Reality, In Proceedings of the 8th International Symposium on Visual Computing (ISVC'12), 16-19. July 2012, Rethymnon, Greece, pp 675-685